原版影印说明

1. 《凝聚态物质与材料数据手册》（6册）是 *Springer Handbook of Condensed Matter and Materials Data* 的影印版。为使用方便，由原版1卷改为6册：

  第1册　通用表和元素

  第2册　材料类：金属材料

  第3册　材料类：非金属材料

  第4册　功能材料：半导体和超导体

  第5册　功能材料：磁性材料、电介质、铁电体和反铁电体

  第6册　特种结构

2. 全书目录、作者信息、缩略语表、索引在各册均完整呈现。

本手册数据全面准确，1 025个图和914个表使查阅更加方便，是非常实用的案头参考书，适于材料及相关专业本科生、研究生、专业研究人员使用。

材料科学与工程图书工作室

  联系电话　0451-86412421

      　0451-86414559

  邮　　　箱　yh_bj@aliyun.com

      　xuyaying81823@gmail.com

      　zhxh6414559@aliyun.com

Springer 手册精选原版系列

凝聚态物质与材料数据手册

材料类：非金属材料

【第3册】

Springer
Handbook of
Condensed Matter
and Materials Data

W.Martienssen

H.Warlimont

Editors

哈尔滨工业大学出版社
HARBIN INSTITUTE OF TECHNOLOGY PRESS

黑版贸审字08-2014-009号

Reprint from English language edition:
Springer Handbook of Condensed Matter and Materials Data
by Werner Martienssen and Hans Warlimont
Copyright © 2005 Springer Berlin Heidelberg
Springer Berlin Heidelberg is a part of Springer Science+Business Media
All Rights Reserved

This reprint has been authorized by Springer Science & Business Media for distribution in China Mainland only and not for export therefrom.

图书在版编目（CIP）数据

凝聚态物质与材料数据手册. 第3册，材料类. 非金属材料：英文 /（德）马蒂安森（Martienssen, W.），（德）沃利蒙特（Warlimont, H.）主编. —哈尔滨：哈尔滨工业大学出版社，2014.3
（Springer手册精选原版系列）
ISBN 978-7-5603-4457-7

Ⅰ. ①凝… Ⅱ. ①马… ②沃… Ⅲ. ①凝聚态–材料–技术手册–英文 ②非金属材料–技术手册–英文 Ⅳ. ①O469-62 ②TB32-62

中国版本图书馆CIP数据核字（2013）第293897号

责任编辑	张秀华　杨　桦　许雅莹
出版发行	哈尔滨工业大学出版社
社　　址	哈尔滨市南岗区复华四道街10号　邮编150006
传　　真	0451-86414749
网　　址	http://hitpress.hit.edu.cn
印　　刷	哈尔滨市石桥印务有限公司
开　　本	787mm×960mm　1/16　印张12.25
版　　次	2014年3月第1版　2014年3月第1次印刷
书　　号	ISBN 978-7-5603-4457-7
定　　价	58.00元

（如因印刷质量问题影响阅读，我社负责调换）

Springer Handbook
of Condensed Matter and Materials Data

W. Martienssen and H. Warlimont (Eds.)

With 1025 Figures and 914 Tables

Springer Handbook provides a concise compilation of approved key information on methods of research, general principles, and functional relationships in physics and engineering. The world's leading experts in the fields of physics and engineering will be assigned by one or several renowned editors to write the chapters comprising each volume. The content is selected by these experts from Springer sources (books, journals, online content) and other systematic and approved recent publications of physical and technical information.

The volumes will be designed to be useful as readable desk reference book to give a fast and comprehensive overview and easy retrieval of essential reliable key information, including tables, graphs, and bibliographies. References to extensive sources are provided.

Preface

The Springer Handbook of Condensed Matter and Materials Data is the realization of a new concept in reference literature, which combines introductory and explanatory texts with a compilation of selected data and functional relationships from the fields of solid-state physics and materials in a single volume. The data have been extracted from various specialized and more comprehensive data sources, in particular the Landolt–Börnstein data collection, as well as more recent publications. This Handbook is designed to be used as a desktop reference book for fast and easy finding of essential information and reliable key data. References to more extensive data sources are provided in each section. The main users of this new Handbook are envisaged to be students, scientists, engineers, and other knowledge-seeking persons interested and engaged in the fields of solid-state sciences and materials technologies.

The editors have striven to find authors for the individual sections who were experienced in the full breadth of their subject field and ready to provide succinct accounts in the form of both descriptive text and representative data. It goes without saying that the sections represent the individual approaches of the authors to their subject and their understanding of this task. Accordingly, the sections vary somewhat in character. While some editorial influence was exercised, the flexibility that we have shown is deliberate. The editors are grateful to all of the authors for their readiness to provide a contribution, and to cooperate in delivering their manuscripts and by accepting essentially all alterations which the editors requested to achieve a reasonably coherent presentation.

An onerous task such as this could not have been completed without encouragement and support from the publisher. Springer has entrusted us with this novel project, and Dr. Hubertus von Riedesel has been a persistent but patient reminder and promoter of our work throughout. Dr. Rainer Poerschke has accompanied and helped the editors constantly with his professional attitude and very personable style during the process of developing the concept, soliciting authors, and dealing with technical matters. In the later stages, Dr. Werner Skolaut became a relentless and hard-working member of our team with his painstaking contribution to technically editing the authors' manuscripts and linking the editors' work with the copy editing and production of the book.

Prof. Werner Martienssen

Prof. Hans Warlimont

We should also like to thank our families for having graciously tolerated the many hours we have spent in working on this publication.

We hope that the users of this Handbook, whose needs we have tried to anticipate, will find it helpful and informative. In view of the novelty of the approach and any possible inadvertent deficiencies which this first edition may contain, we shall be grateful for any criticisms and suggestions which could help to improve subsequent editions so that they will serve the expectations of the users even better and more completely.

September 2004
Frankfurt am Main, Dresden

Werner Martienssen,
Hans Warlimont

About the Authors

Wolf Assmus

Johann Wolfgang Goethe-University
Physics Department
Frankfurt am Main, Germany
assmus@physik.uni-frankfurt.de
http://www.rz.uni-frankfurt.de/piweb/
kmlab/Leiter.html

Chapter 1.3

Dr. Wolf Assmus (Kucera Professor) is Professor of Physics at the University of Frankfurt and Dean of the Physics-Faculty. He is a solid state physicist, especially interested in materials research and crystal growth. His main research fields are: materials with high electronic correlation, quasicrystals, materials with extremely high melting temperatures, magnetism, and superconductivity.

Stefan Brühne

Johann Wolfgang Goethe-University
Physics Department
Frankfurt am Main, Germany
bruehne@physik.uni-frankfurt.de

Chapter 1.3

Dr. Stefan Brühne, née Mahne, a chemist by education in Germany and England, received his PhD in 1994 from Dortmund University, Germany, on giant cell crystal structures in the Al–Ta system. Following a post doc position at the Materials Department (Crystallography) at ETH Zurich he spent seven years in the ceramics industry. His main activity was R&D of glasses, frits and pigments for high-temperature applications, thereby establishing design of experiment (DoE) techniques. Since 2002, at the Institute of Physics at Frankfurt University he has been investigated X-ray structure determination of quasicrystalline, highly complex and disordered intermetallic materials.

Fabrice Charra

Commissariat à l'Énergie Atomique,
Saclay
Département de Recherche sur l'État
Condensé, les Atomes et les Molécules
Gif-sur-Yvette, France
fabrice.charra@cea.fr
http://www-drecam.cea.fr/spcsi/

Chapter 5.3

Fabrice Charra conducts research in the emerging field of nanophotonics, in the surface physics laboratory of CEA/Saclay. The emphasis of his work is on light emission and absorption form single nanoscale molecular systems. His area of expertise also extends to nonlinear optics, a domain to which he contributed several advances in the applications of organic materials.

Gianfranco Chiarotti

University of Rome "Tor Vergata"
Department of Physics
Roma, Italy
chiarotti@roma2.infn.it

Chapter 5.2

Gianfranco Chiarotti is Professor Emeritus, formerly Professor of General Physics, Fellow of the American Physical Society, fellow of the Italian National Academy (Accademia Nazionale dei Lincei). He was Chairman of the Physics Committee of the National Research Council (1988–1994), Chair Franqui at the University of Liège (1975), Assistant Professor at the University of Illinois (1955–1957), Editor of the journal Physics of Solid Surfaces, and Landolt–Börnstein Editor of Springer-Verlag from 1993 through 1996. He has worked in several fields of solid state physics, namely electronic properties of defects, modulation spectroscopy, optical properties of semiconductors, surface physics, and scanning tunnelling microscopy (STM) in organic materials.

Claus Fischer

Formerly Institute of Solid State and
Materials Research (IFW)
Dresden, Germany
A_C.FischerDD@t-online.de

Chapter 4.2

Claus Fischer received his PhD from the Technical University Dresden (Since his retirement in 2000 he continues to work as a foreign scientist of IFW in the field of high-T_c superconductors.) His last position at IFW was head of the Department of Superconducting Materials. The main areas of research were growth of metallic single crystals in particular of magnetic materials, developments of hard magnetic materials, of materials for thick film components of microelectronics and of low-T_c and high-T_c superconducting wires and tapes. Many activities were performed in cooperation with industrial manufacturers.

About the Authors

Günter Fuchs
Leibniz Institute for Solid State and
Materials Research (IFW) Dresden
Magnetism and Superconductivity in the
Institute of Metallic Materials
Dresden, Germany
fuchs@ifw-dresden.de
http://www.ifw-dresden.de/imw/21/

Chapter 4.2

Dr. Günter Fuchs studied physics at the Technical University of Dresden, Germany, and received his PhD in 1980 on the pinning mechanism in superconducting NbTi alloys. Since 1969 he has been at the Institute of Solid State and Materials Research (IFW) in Dresden. His activities are in superconductivity (HTSC, MgB_2, intermetallic borocarbides) and the applications of superconductors. He received the PASREG Award for outstanding scientific achievements in the field of bulk cuprate superconductors in high magnetic fields in 2003.

Frank Goodwin
International Lead Zinc Research
Organization, Inc.
Research Triangle Parc, NC, USA
fgoodwin@ilzro.org
http://www.ilzro.org/Contactus.htm

Chapter 3.1

Frank Goodwin received his Sc.D. from the Massachusetts Institute of Technology in 1979 and is responsible for all materials science research at International Lead Zinc Research Organization, Inc. where he has conceived and managed numerous projects on lead and zinc-containing products. These have included lead in acoustics, cable sheathing, nuclear waste management and specialty applications, together with zinc in coatings, castings and wrought forms.

Susana Gota-Goldmann
Commissariat à l'Energie Atomique (CEA)
Direction de la Recherche Technologique (DRT)
Fontenay aux Roses, France
susana.gota-goldmann@cea.fr

Chapter 5.3

Dr. Susana Gota-Goldmann received her PhD in Materials Science form the Université Pierre et Marie Curie (Paris V) in 1993. After her PhD, she was engaged as a researcher in the Materials Science Division of the CEA (Commissariat à l'Energie Atomique, France). She has focused her scientific activity on the growth and characterisation of nanometric oxide layers with applications in spin electronics and photovoltaics. In parallel she has developed the use of synchrotron radiation techniques (X-ray absorption magnetic dicroism, photoemission, resonant reflectivity) for the study of oxide thin layers. Recently she has moved from fundamental to technological research. Dr. Gota-Goldmann is now working as a project manager at the scientific affairs direction of the Technology Research Division (CEA/DRT).

Sivaraman Guruswamy
University of Utah
Metallurgical Engineering
Salt Lake City, UT, USA
sguruswa@mines.utah.edu
http://www.mines.utah.edu/metallurgy/MML

Chapter 3.1

Dr. Guruswamy is a Professor of Metallurgical Engineering at the University of Utah. He obtained his Ph.D. degree in Metallurgical Engineering from the Ohio State University in 1984. He has made significant contributions in several areas including magnetic materials development, deformation of compound semiconductors, and lead alloys. His current work focuses on magnetostrictive materials and hybrid thermionic/thermoelectric thermal diodes.

Gagik G. Gurzadyan
Technical University of Munich
Institute for Physical and Theoretical
Chemistry
Garching, Germany
gurzadyan@ch.tum.de
http://zentrum.phys.chemie.
tu-muenchen.de/gagik

Chapter 4.4

Gagik G. Gurzadyan, Ph.D., Dr. Sci., has extensive experience in nonlinear optics and crystals, laser photophysics and spectroscopy. He has authored several books including the Handbook of Nonlinear Optical Crystals published by Springer-Verlag. He worked in the Institute of Spectroscopy (USSR), CEA/Saclay (France), Max-Planck-Institute of Radiation Chemistry (Germany). At present he works at the Technical University of Munich with ultrafast lasers in the fields of nonlinear photochemistry of biomolecules and femtosecond spectroscopy.

About the Authors

Hideki Harada

High Tech Association Ltd.
Higashikaya, Fukaya, Saitama, Japan
khb16457@nifty.com
http://homepage1.nifty.com/JABM

Chapter 4.3

Dr. Hideki Harada is chief advisor of magnetic materials and their application and President of High Tech Association Ltd., Saitama, Japan. He is Chairman of the Japan Association of Bonded Magnet Industries (JABM) and received his Ph.D. in 1987 with a work on electrostatic ferrite materials. He worked in research and development of magnetic materials and cemented carbide tools at Hitachi Metals where he also was on the Board of Directors. He received the Japanese National Award for Industries Development Contribution.

Bernhard Holzapfel

Leibniz Institute for Solid State and Materials Research Dresden – Institute of Metallic Materials
Superconducting Materials
Dresden, Germany
B.Holzapfel@ifw-dresden.de
http://www.ifw-dresden.de/imw/26/

Chapter 4.2

Dr. Bernhard Holzapfel is head of the superconducting materials group at the Leibniz Institute for Solid State and Materials Research (IFW) Dresden, Germany. His main area of research is pulsed laser deposition of functional thin films and superconductivity. Currently he works on the development of HTSC high J_c coated conductors using ion beam assisted deposition or highly textured metal substrates. His work is supported by a number of national and European founded research projects.

Karl U. Kainer

GKSS Research Center Geesthacht
Institute for Materials Research
Geesthacht, Germany
karl.kainer@gkss.de
http://www.gkss.de

Chapter 3.1

Professor Kainer is director of Institute for Materials Research at GKSS-Research Center, Geesthacht and Professor of Materials Technology at the Technical University of Hamburg-Harburg. He obtained his Ph.D. in Materials Science at the Technical University of Clausthal in 1985 and his Habilitation in 1996. In 1988 he received the Japanese Government Research Award for Foreign Specialists. His current research activities are the development of new alloys and processes for magnesium materials.

Catrin Kammer

METALL – Intl. Journal for Metallurgy
Goslar, Germany
Kammer@metall-news.com
http://www.giesel-verlag.de

Chapter 3.1

Catrin Kammer received her Ph.D. in materials sciences from the Technical University Bergakademie Freiberg, Germany, in 1989. She has been working in the field of light metals and is author of several handbooks about aluminium and magnesium. She is working as author for the journal ALUMINIUM and is teaching in material sciences. Since 2001 she is editor-in-chief of the journal METALL, which deals with all non-ferrous metals.

Wolfram Knabl

Plansee AG
Technology Center
Reutte, Austria
wolfram.knabl@plansee.com
http://www.plansee.com

Chapter 3.1

Dr. Wolfram Knabl studied materials science at the Mining University of Leoben, Austria and received his Ph.D. at the Plansee AG focusing on the development of oxidation protective coatings for refractory metals. Between 1996 and 2002 he was responsible for the test laboratories at Plansee AG and since October 2002 he is working in the field of refractory metals, especially material and process development in the technology center of Plansee AG.

About the Authors

Alfred Koethe

Leibniz-Institut für Festkörper- und Werkstoffforschung
Institut für Metallische Werkstoffe (retired)
Dresden, Germany
alfred.koethe@web.de

Chapter 3.1

Dr. Alfred Koethe is physicist and professor of Materials Science. He retired in 2000 from his position as head of department in the Institute of Metallic Materials at the Leibniz Institute of Solid State and Materials Research in Dresden, Germany. His main research activities were in the fields of preparation and properties of ultrahigh-purity refractory metals and, especially, of steels (stainless steels, high strenght steels, thermomechanical treatment, microalloying, relations chemical composition/microstructure/properties).

Dieter Krause

Schott AG
Research and Technology-Development
Mainz, Germany
dieter.krause@schott.com

Chapter 3.4

Dieter Krause studied physics at the universities of Erlangen and Munich, Germany, where he received his Ph.D. for work on magnetism and metal physics. He was professor in Tehran, Iran, lecturer in Munich and Mainz, Germany. As scientist and director of Schott's corporate research and development centre he was involved in research on optical and mechanical properties of amorphous materials, thin films, and optical fibres. Now he is consultant, chief scientist, and the editor of the "Schott Series on Glass and Glass Ceramics – Science, Technology, and Applications" published by Springer.

Manfred D. Lechner

Universität Osnabrück
Institut für Chemie – Physikalische Chemie
Osnabrück, Germany
lechner@uni-osnabrueck.de
http://www.chemie.uni-osnabrueck.de/pc/index.html

Chapter 3.3

Professor Lechner has a PhD in chemistry from the University of Mainz, Germany. Since 1975 he is Professor of Physical Chemistry at the Institute of Chemistry of the University of Osnabrück, Germany. His scientific work concentrates on the physics and chemistry of polymers. In this area he is mainly working on the influence of high pressure on polymer systems, polymers for optical storage and waveguides as well as synthesis and properties of superabsorbers from renewable resources.

Gerhard Leichtfried

Plansee AG
Technology Center
Reutte, Austria
gerhard.leichtfried@plansee.com
http://www.plansee.com

Chapter 3.1

Dr. Gerhard Leichtfried received his Ph.D from the Montanuniversität Leoben and is qualified for lecturing in powder metallurgy. For 20 years he has been working in various senior positions for the Plansee Aktiengesellschaft, a company engaged in refractory metals, composite materials, cemented carbides and sintered iron and steels.

Werner Martienssen

Universität Frankfurt/Main
Physikalisches Institut
Frankfurt/Main, Germany
Martienssen@Physik.uni-frankfurt.de

Chapters 1.1, 1.2, 2.1, 4.1

Werner Martienssen studied physics and chemistry at the Universities of Würzburg and Göttingen, Germany. He obtained his Ph.D. in Physics with R.W. Pohl, Göttingen, and holds an honorary doctorate at the University of Dortmund. After a visiting-professorship at the Cornell University, Ithaca, USA in 1959 to 1960 he taught physics at the University of Stuttgart and since 1961 at the University of Frankfurt/Main. His main research fields are condensed matter physics, quantum optics and chaotic dynamics. Two of his former students and coworkers became Nobel-laureates in Physics, Gerd K. Binnig for the design of the scanning tunneling microscope in 1986 and Horst L. Störmer for the discovery of a new form of quantum-fluid with fractionally charged excitations in 1998. Werner Martienssen is a member of the Deutsche Akademie der Naturforscher Leopoldina, Halle and of the Akademie der Wissenschaften zu Göttingen. Since 1994 he is Editor-in-Chief of the data collection Landolt–Börnstein published by Springer, Heidelberg.

About the Authors

Toshio Mitsui

Osaka University
Takarazuka, Japan
t-mitsui@jttk.zaq.ne.jp

Chapter 4.5

Toshio Mitsui is an emeritus professor of Osaka University. He studied solid state physics and biophysics at Hokkaido University, Pennsylvania State University, Brookhaven National Laboratory, the Massachusetts Institute of Technology, Osaka University and Meiji University. He was the first to observe the ferroelectric domain structure in Rochelle salt with a polarization microscope. He proposed various theories on ferroelectric effects and biological molecular machines.

Manfred Müller

Dresden University of Technology
Institute of Materials Science
Dresden, Germany
m.mueller33@t-online.de

Chapter 4.3

Dr.-Ing. habil. Manfred Müller is a Professor emeritus of Special Materials at the Institute of Materials Science of the Dresden University of Technology. Before his retirement he was for many years head of department for special materials at the Central Institute for Solid State Physics and Materials Research of the Academy of Sciences in Dresden, Germany. His main field was the research and development of metallic materials with emphasis on special physical properties, such as soft and hard magnetic, electrical and thermoelastic properties. His last field of research was amorphous and nanocrystalline soft magnetic alloys. He is a member of the German Society of Materials Science (DGM) and was a member of the Advisory Board of DGM.

Sergei Pestov

Moscow State Academy of Fine Chemical Technology
Department of Inorganic Chemistry
Moscow, Russia
pestovsm@yandex.ru

Chapter 5.1

Dr. Pestov is a docent of the Inorganic Chemistry Department and a head of group on liquid crystals (LC) at the Moscow State Academy of Fine Chemical Technology. He earned his Ph.D. in physical chemistry in 1992. His research is focused on thermal analysis and thermodynamics of systems containing LC and physical properties of LC. He is an author of a Landolt–Börnstein volume and two books devoted to liquid crystals.

Günther Schlamp

Metallgesellschaft Ffm and Degussa Demetron (retired)
Steinbach/Ts, Germany

Chapter 3.1

Günther Schlamp received his Ph.D. from the Johann-Wolfgang-Goethe University of Frankfurt/Main, Germany, in Physical Chemistry. His industrial activities in research include the development and production of refractory material coatings, high purity materials and parts for electronics, and sputter targets for the reflection-enhancing coating of glas. He has contributed to several Handbooks with repoprts on properties and applications of noble metals and their alloys.

Barbara Schüpp-Niewa

Leibniz-Institute for Solid State and Materials Research Dresden
Institute for Metallic Materials
Dresden, Germany
b.schuepp@ifw-dresden.de
http://www.ifw-dresden.de

Chapter 4.2

Barbara Schüpp-Niewa studied chemistry in Gießen and Dortmund where she received her Ph.D. in 1999. Since 2000 she has been a scientist at the Leibniz-Institute for Solid State and Materials Research Dresden with a focus on crystal structure investigations of oxometalates with superconducting or exciting magnetic ground states. Her current research activities include coated conductors.

About the Authors

Roland Stickler

University of Vienna
Department of Chemistry
Vienna, Austria
roland.stickler@univie.ac.at

Chapter 3.1

Professor Stickler received his master and Dr. degree from the Technical University in Vienna. From 1958 to 1972 he was manager of physical metallurgy with the Westinghouse Research Laboratory in Pittsburgh, Pa. In 1972 he accepted a full professorship at the University of Vienna heading a materials science group in the Institute of Physical Chemistry, and from 1988 he was head of this institute until his retirement as professor emeritus in 1998. He was involved in research and engineering work on superalloys, semiconductor materials and high melting point materials, investigating the relationship between microstructure and mechanical behavior, in particular fatigue and fracture mechanics properties. He was leader of a successful project on brazing under microgravity conditions in the Spacelab-Mission. Further activities included the participation in European COST projects, in particular as chairman of actions on powder metallurgy and light metals. He has authored and coauthored more than 250 publications in scientific journals and proceedings.

Pancho Tzankov

Max Born Institute for Nonlinear Optics
and Short Pulse Spectroscopy
Berlin, Germany
tzankov@mbi-berlin.de
http://staff.mbi-berlin.de/tzankov/

Chapter 4.4

Pancho Tzankov studied laser physics at Sofia University, Bulgaria, and received his Ph.D. in physical chemistry from the Technical University of Munich, Germany. He is now a postdoctoral fellow at the Max Born Institute in Berlin, Germany. His research activities involve development of new nonlinear optical parametric sources of ultrashort pulses and their application for time-resolved spectroscopy.

Volkmar Vill

University of Hamburg
Department of Chemistry, Institute of
Organic Chemistry
Hamburg, Germany
vill@chemie.uni-hamburg.de
http://liqcryst.chemie.uni-hamburg.de/

Chapter 5.1

Professor Volkmar Vill received his Diploma in Chemistry in 1986, his Diploma in Physics in 1988 and his Ph.D. in Chemistry in 1990 from the University of Münster, Germany. In 1997 he earned his Habilitation in Organic Chemistry from the University of Hamburg where he is Professor of Organic Chemistry since 2002. He is the author of the LiqCryst – Database of Liquid Crystals and the Editor of the Handbook of Liquid Crystals, of Landolt–Börnstein, Organic Index, and Vol. VIII/5a, Physical Properties of Liquid Crystals.

Hans Warlimont

DSL Dresden Material-Innovation GmbH
Dresden, Germany
warlimont@ifw-dresden.de

Chapters 3.1, 3.2, 4.2, 4.3

Hans Warlimont is a physical metallurgist and has worked on numerous topics in several research institutions and industrial companies. Among them were the Max-Planck-Institute of Metals Research, Stuttgart, and Vacuumschmelze, Hanau. He was Scientific Director of the Leibniz-Institute of Solid State and Materials Research Dresden and Professor of Materials Science at Dresden University of Technology. Recently he has established DSL Dresden Material-Innovation GmbH to industrialise his invention of electroformed battery grids.

Acknowledgements

2.1 The Elements
by Werner Martienssen

We thank Dr. G. Leichtfried, Plansee AG, A-6600 Reutte/Tirol for recently determined new data on the refractory metals Nb, Ta, and Mo, W.

4.1 Semiconductors
by Werner Martienssen

In selecting the "most important information" from the huge data collection in Landolt–Börnstein, the author found great help in the new *Semiconductors: Data Handbook* [1]. Again, the data in this Springer Handbook of Condensed Matter and Materials Data represent only a small fraction of the information given in *Semiconductors: Data Handbook*, which is about 700 pages long. I am much indebted to my colleague O. Madelung for kindly presenting me the manuscript of that Handbook prior to publication.

[1] O. Madelung (Ed.): *Semiconductors: Data Handbook*, 3rd Edn. (Springer, Berlin, Heidelberg 2004)

4.5 Ferroelectrics and Antiferroelectrics
by Toshio Mitsui

The author of this subchapter thanks the coauthors of LB III/36 for their helpful discussions and suggestions. Especially, he is much indebted to Prof. K. Deguchi for his kind support throughout the preparation of the manuscript.

Contents

| List of Abbreviations | 19 |

第1册 通用表和元素

Part 1 General Tables

1 The Fundamental Constants
Werner Martienssen 3
   1.1 What are the Fundamental Constants
        and Who Takes Care of Them? 3
   1.2 The CODATA Recommended Values of the Fundamental Constants 4
   References 9

2 The International System of Units (SI), Physical Quantities, and Their Dimensions
Werner Martienssen 11
   2.1 The International System of Units (SI) 11
   2.2 Physical Quantities 12
   2.3 The SI Base Units 13
   2.4 The SI Derived Units 16
   2.5 Decimal Multiples and Submultiples of SI Units 19
   2.6 Units Outside the SI 20
   2.7 Some Energy Equivalents 24
   References 25

3 Rudiments of Crystallography
Wolf Assmus, Stefan Brühne 27
   3.1 Crystalline Materials 28
   3.2 Disorder 38
   3.3 Amorphous Materials 39
   3.4 Methods for Investigating Crystallographic Structure 39
   References 41

Part 2 The Elements

1 The Elements
Werner Martienssen 45
   1.1 Introduction 45
   1.2 Description of Properties Tabulated 46
   1.3 Sources 49
   1.4 Tables of the Elements in Different Orders 49
   1.5 Data 54
   References 158

第2册 材料类：金属材料

Part 3 Classes of Materials

1 Metals
Frank Goodwin, Sivaraman Guruswamy, Karl U. Kainer, Catrin Kammer, Wolfram Knabl, Alfred Koethe, Gerhard Leichtfried, Günther Schlamp, Roland Stickler, Hans Warlimont 161
- 1.1   Magnesium and Magnesium Alloys 162
- 1.2   Aluminium and Aluminium Alloys 171
- 1.3   Titanium and Titanium Alloys 206
- 1.4   Zirconium and Zirconium Alloys 217
- 1.5   Iron and Steels 221
- 1.6   Cobalt and Cobalt Alloys 272
- 1.7   Nickel and Nickel Alloys 279
- 1.8   Copper and Copper Alloys 296
- 1.9   Refractory Metals and Alloys 303
- 1.10 Noble Metals and Noble Metal Alloys 329
- 1.11 Lead and Lead Alloys 407
- References 422

第3册 材料类：非金属材料（本册）

2 Ceramics
Hans Warlimont 431
- 2.1   Traditional Ceramics and Cements 432
- 2.2   Silicate Ceramics 433
- 2.3   Refractory Ceramics 437
- 2.4   Oxide Ceramics 437
- 2.5   Non-Oxide Ceramics 451
- References 476

3 Polymers
Manfred D. Lechner 477
- 3.1   Structural Units of Polymers 480
- 3.2   Abbreviations 482
- 3.3   Tables and Figures 483
- References 522

4 Glasses
Dieter Krause 523
- 4.1   Properties of Glasses – General Comments 526
- 4.2   Composition and Properties of Glasses 527
- 4.3   Flat Glass and Hollowware 528
- 4.4   Technical Specialty Glasses 530
- 4.5   Optical Glasses 543
- 4.6   Vitreous Silica 556
- 4.7   Glass-Ceramics 558

	4.8	Glasses for Miscellaneous Applications	559
		References	572

第4册 功能材料：半导体和超导体

Part 4 Functional Materials

1 Semiconductors
Werner Martienssen 575
	1.1	Group IV Semiconductors and IV–IV Compounds	578
	1.2	III–V Compounds	604
	1.3	II–VI Compounds	652
		References	691

2 Superconductors
Claus Fischer, Günter Fuchs, Bernhard Holzapfel, Barbara Schüpp-Niewa, Hans Warlimont 695
	2.1	Metallic Superconductors	696
	2.2	Non-Metallic Superconductors	711
		References	749

第5册 功能材料：磁性材料、电介质、铁电体和反铁电体

3 Magnetic Materials
Hideki Harada, Manfred Müller, Hans Warlimont 755
	3.1	Basic Magnetic Properties	755
	3.2	Soft Magnetic Alloys	758
	3.3	Hard Magnetic Alloys	794
	3.4	Magnetic Oxides	811
		References	814

4 Dielectrics and Electrooptics
Gagik G. Gurzadyan, Pancho Tzankov 817
	4.1	Dielectric Materials: Low-Frequency Properties	822
	4.2	Optical Materials: High-Frequency Properties	824
	4.3	Guidelines for Use of Tables	826
	4.4	Tables of Numerical Data for Dielectrics and Electrooptics	828
		References	890

5 Ferroelectrics and Antiferroelectrics
Toshio Mitsui 903
	5.1	Definition of Ferroelectrics and Antiferroelectrics	903
	5.2	Survey of Research on Ferroelectrics	904
	5.3	Classification of Ferroelectrics	906
	5.4	Physical Properties of 43 Representative Ferroelectrics	912
		References	936

第6册 特种结构

Part 5 Special Structures

1 Liquid Crystals
Sergei Pestov, Volkmar Vill 941
  1.1 Liquid Crystalline State 941
  1.2 Physical Properties of the Most Common Liquid Crystalline Substances 946
  1.3 Physical Properties of Some Liquid Crystalline Mixtures 975
  References.. 977

2 The Physics of Solid Surfaces
Gianfranco Chiarotti 979
  2.1 The Structure of Ideal Surfaces 979
  2.2 Surface Reconstruction and Relaxation 986
  2.3 Electronic Structure of Surfaces 996
  2.4 Surface Phonons 1012
  2.5 The Space Charge Layer at the Surface of a Semiconductor 1020
  2.6 Most Frequently Used Acronyms 1026
  References.. 1029

3 Mesoscopic and Nanostructured Materials
Fabrice Charra, Susana Gota-Goldmann 1031
  3.1 Introduction and Survey 1031
  3.2 Electronic Structure and Spectroscopy... 1035
  3.3 Electromagnetic Confinement 1044
  3.4 Magnetic Nanostructures 1048
  3.5 Preparation Techniques 1063
  References.. 1066

Subject Index
Periodic Table of the Elements
Most Frequently Used Fundamental Constants

List of Abbreviations

2D-BZ	2-dimensional Brillouin zone
2P-PES	2-photon photoemission spectroscopy

A

AES	Auger electron spectroscopy
AFM	atomic force microscope
AISI	American Iron and Steel Institute
APS	appearance potential spectroscopy
ARUPS	angle-resolved ultraviolet photoemission spectroscopy
ARXPS	angle-resolved X-ray photoemission spectroscopy
ASTM	American Society for Testing and Materials
ATR	attenuated total reflection

B

BBZ	bulk Brillouin zone
BIPM	Bureau International des Poids et Mesures
BZ	Brillouin zone

C

CB	conduction band
CBM	conduction band minimum
CISS	collision ion scattering spectroscopy
CITS	current imaging tunneling spectroscopy
CMOS	complementary metal–oxide–semiconductor
CODATA	Committee on Data for Science and Technology
CVD	chemical vapour deposition

D

DFB	distributed-feedback
DFG	difference frequency generation
DOS	density of states
DSC	differential scanning calorimetry
DTA	differential thermal analysis

E

EB	electron-beam melting
ECS	electron capture spectroscopy
EELS	electron-energy loss spectroscopy
ELEED	elastic low-energy electron diffraction
ESD	electron-stimulated desorption
EXAFS	extended X-ray absorption fine structure

F

FEM	field emission microscope/microscopy
FIM	field ion microscope/microscopy

G

GMR	giant magnetoresistance

H

HAS	helium atom scattering
HATOF	helium atom time-of-flight spectroscopy
HB	Brinell hardness number
HEED	high-energy electron diffraction
HEIS	high-energy ion scattering/high-energy ion scattering spectroscopy
HK	Knoop hardness
HOPG	highly oriented pyrolytic graphite
HPDC	high-pressure die casting
HR-EELS	high-resolution electron energy loss spectroscopy
HR-LEED	high-resolution LEED
HR-RHEED	high-resolution RHEED
HREELS	high-resolution electron energy loss spectroscopy
HRTEM	high-resolution transition electron microscopy
HT	high temperature
HTSC	high-temperature superconductor
HV	Vicker's Hardness

I

IACS	International Annealed Copper Standard
IB	ion bombardment
IBAD	ion-beam-assisted deposition
ICISS	impact ion scattering spectroscopy
ICSU	International Council of the Scientific Unions
IPE	inverse photoemission
IPES	inverse photoemission spectroscopy
ISO	International Organization for Standardization
ISS	ion scattering spectroscopy
IUPAC	International Union of Pure and Applied Chemistry

J

JDOS	joint density of states

K

KRIPES	K-resolved inverse photoelectron spectroscopy

L

LAPW	linearized augmented-plane-wave method
LB	Langmuir–Blodgett
LCM	liquid crystal material
LCP	liquid crystal polymer
LCs	liquid crystals
LDA	local-density approximation
LDOS	local density of states
LEED	low-energy electron diffraction
LEIS	low-energy ion scattering/low-energy ion scattering spectroscopy
LPE	liquid phase epitaxy

M

MBE	molecular-beam epitaxy
MD	molecular dynamics
MEED	medium-energy electron diffraction
MEIS	medium-energy ion scattering/medium-energy ion scattering spectroscopy
MFM	magnetic force microscopy
ML	monolayer
MOCVD	metal-organic chemical vapor deposition
MOKE	magneto-optical Kerr effect
MOSFET	MOS field-effect transistor
MQW	multiple quantum well

N

NICISS	neutral impact collision ion scattering spectroscopy
NIMs	National Institutes for Metrology

O

OPO	optical parametric oscillation

P

PDS	photothermal displacement spectroscopy
PED	photoelectron diffraction
PES	photoemission spectroscopy
PLAP	pulsed laser atom probe
PLD	pulsed laser deposition
PSZ	stabilized zirconia
PZT	piezoelectric material

R

RAS	reflectance anisotropy spectroscopy
RE	rare earth
REM	reflection electron microscope/microscopy
RHEED	reflection high-energy electron diffraction
RIE	reactive ion etching
RPA	random-phase approximation
RT	room temperature
RTP	room temperaure and standard pressure

S

SAM	self-assembled monolayer
SAM	scanning Auger microscope/microscopy
SARS	scattering and recoiling ion spectroscopy
SAW	surface acoustic wave
SBZ	surface Brillouin zone
SCLS	surface core level shift
SDR	surface differential reflectivity
SEM	scanning electron microscope
SEXAFS	surface-sensitive EXAFS
SFG	sum frequency generation
SH	second harmonic
SHG	second-harmonic generation
SI	Système International d'Unités
SIMS	secondary-ion mass spectroscopy
SNR	signal-to-noise ratio
SPARPES	spin polarized angle-resolved photoemission spectroscopy
SPIPES	spin-polarized inverse photoemission spectroscopy
SPLEED	spin-polarized
SPV	surface photovoltage spectroscopy
SQUIDS	superconducting quantum interference devices
SS	surface state
STM	scanning tunneling microscope/microscopy
STS	scanning tunneling spectroscopy
SXRD	surface X-ray diffraction

T

TAFF	thermally activated flux flow
TEM	transmission electron microscope/microscopy
TFT	thin-film transistor
TMR	tunnel magnetoresistance
TMT	thermomechanical treatment
TOF	time of flight
TOM	torsion oscillation magnetometry

TRS	truncation rod scattering	**V**	
TTT	time-temperature-transformation	VBM	valence band maximum
U		VLEED	very low-energy electron diffraction
UHV	ultra-high vacuum	**X**	
UPS	ultraviolet photoemission spectroscopy		
UV	ultraviolet	XPS	X-ray photoemission spectroscopy

3.2. Ceramics

Ceramics have various definitions, because of their long history of development as one of the oldest and most versatile groups of materials and because of the different ways in which materials can be classified, such as by chemical composition (silicates, oxides and non-oxides), properties (mechanical and physical), or applications (building materials, high-temperature materials and functional materials). The most widely used, minimal definition of ceramics is that they are inorganic nonmetallic materials. The broadest subdivision is into traditional, and technical or engineering ceramics. Detailed groupings and definitions of technical ceramics are given in DIN EN 60672. In the present section we differentiate between traditional ceramics and cements, silicate ceramics, refractory ceramics, oxide ceramics, and non-oxide ceramics, being aware that there are overlaps. It should also be noted that other sections of this Handbook cover particular groups of ceramics: glasses (Chapt. 3.4), semiconductors (Chapt. 4.1), nonmetallic superconductors (Sect. 4.2.2), magnetic oxides (Sect. 4.3.4), dielectrics and electrooptics (Chapt. 4.4) and ferroelectrics and related materials, (Chapt. 4.5).

3.2.1 **Traditional Ceramics and Cements**........ 432
    3.2.1.1 Traditional Ceramics 432
    3.2.1.2 Cements................................. 432
3.2.2 **Silicate Ceramics**................................. 433
3.2.3 **Refractory Ceramics** 437
3.2.4 **Oxide Ceramics**................................... 437
    3.2.4.1 Magnesium Oxide.................... 444
    3.2.4.2 Alumina................................. 445
    3.2.4.3 Al–O–N Ceramics..................... 447
    3.2.4.4 Beryllium Oxide 447
    3.2.4.5 Zirconium Dioxide 447
    3.2.4.6 Titanium Dioxide, Titanates, etc. 450
3.2.5 **Non-Oxide Ceramics**........................... 451
    3.2.5.1 Non-Oxide High-Temperature Ceramics 451
    3.2.5.2 Borides 451
    3.2.5.3 Carbides................................. 458
    3.2.5.4 Nitrides.................................. 467
    3.2.5.5 Silicides 473
References 476

Detailed treatments of ceramics are given in [2.1–3]. Reference [2.4] is a comprehensive handbook on materials, emphasizing ceramics and minerals. Structural ceramics are treated in [2.5]. Reference [2.6] is a hands-on practical reference book for technical ceramics, and recent data on technical ceramics can be found in conference proceedings such as [2.7].

3.2.1 Traditional Ceramics and Cements

3.2.1.1 Traditional Ceramics

Traditional ceramics are obtained by the firing of clay-based materials. They are commonly composed of a clay mineral (kaolinite, montmorillonite, or illite), fluxing agents (orthoclase and plagioclase), and filler materials (SiO_2, Al_2O_3, and MgO). The processing steps are mixing, forming, drying, firing (high-temperature treatment in air), and finishing (enameling, cleaning, and machining). The main classes of traditional ceramics are fired bricks, whiteware (china, stoneware, and porcelain), glazes, porcelain enamels, high-temperature refractories, cements, mortars, and concretes. Table 3.2-1 lists some of main groups of traditional ceramics, with some of their properties and applications.

3.2.1.2 Cements

Cement is the common binder of traditional building ceramics. It is produced by mixing about 80 wt% of low-magnesium (< 3 wt% MgO) calcium carbonate ($CaCO_3$) (limestone, marl, or chalk) with about 20 wt% clay (which may be obtained from clays, shale, or slag). In terms of oxide content, this corresponds to a ratio of CaO to SiO_2 of 3:1 by weight. The common term "Portland cement" is based on the early use of a particular limestone called Portland stone. The processing steps are milling and mixing, heating to 260 °C, precalcining at 900 °C, and calcining in a rotary kiln at temperatures ≤ 1450 °C. The resulting product is termed "clinker" and consists of a vitreous nodular material composed of calcium silicates and aluminates. This is mixed with 2 to 4 wt% gypsum ($CaSO_4 \cdot 2H_2O$) to adjust the setting time, and ground to the final product. The ranges of the percentages of the constituents are given in Table 3.2-2.

Some standardized grades of Portland cement and their uses are listed in Table 3.2-3. Mortars and concretes are mixtures of cement with specified amounts of sand, gravel, or crushed stones with specified particle sizes.

Table 3.2-2 Chemical composition of Portland cement [2.4]

Component	Average mass fraction (wt%)
SiO_2	21.8–21.9
Al_2O_3	4.9–6.9
Fe_2O_3	2.4–2.9
CaO	63.0–65.0
MgO	1.1–2.5 (max. 3.0)
SO_3	1.7–2.6
Na_2O	0.2
K_2O	0.4
H_2O	1.4–1.5

Table 3.2-1 Examples of traditional ceramics [2.4]

Type	Properties	Applications
Fired brick	Porosity 15–30% Firing temperature 950–1050 °C May be enameled or not	Bricks, pipes, ducts, walls, floor tiles
China	Porosity 10–15% Firing temperature 950–1200 °C Enameled, opaque	Sanitary, tiles
Stoneware	Porosity 0.5–3% Firing temperature 1100–1300 °C Glassy surface	Crucibles, labware, pipes
Porcelain	Porosity 0–2% Firing temperature 1100–1400 °C Glassy, translucent	Insulators, labware, cookware

Table 3.2-3 ASTM Portland cement types [2.4]

ASTM type	Name	Compressive strength after 28 days (MPa)	Applications
Type I	Normal or ordinary Portland cement (NPC)	42	General uses, and hence used when no special properties are required.
Type II	Modified Portland cement (MPC)	47	Low heat generation during the hydration process. Most useful in structures with large cross sections and for drainage pipes where sulfate levels are low.
Type III	Rapid-hardening Portland cement (RHPC)	52	Used when high strength is required after a short period of curing.
Type IV	Low-heat Portland cement (LHPC)	34	Less heat generation during hydration than for Type II. Used for mass concrete construction where large heat generation could create problems. The tricalcium aluminate content must be maintained below 7 wt%.
Type V	Sulfate-resisting Portland cement (SRPC)	41	Has high sulfate resistance. It is a special cement used when severe attack is possible.

3.2.2 Silicate Ceramics

Silicates are the salts or esters of orthosilicic acid (H_4SiO_4) and of its condensation products. The silicates are categorized according to the arrangement of the [SiO_4] tetrahedra in their crystal structure. Figure 3.2-1 shows some simple, planar arrangements. In the tectosilicates, the array of the SiO_4 tetrahedra is three-dimensional. Typical examples are talc, feldspar, and the zeolites.

As silicates are the most important constituents of the earth's crust, they became the basis of the traditional

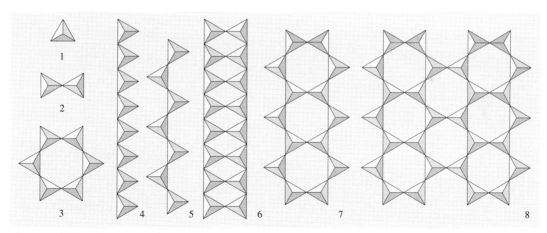

Fig. 3.2-1 Schematic representation of the arrangement of the [SiO_4] tetrahedra in planar silicate crystal structures: 1, nesosilicate; 2, sorosilicate; 3, cyclosilicate; 4 and 5, inosilicates; 6 and 7, ribbon silicates; 8, layered silicate or phyllosilicate

ceramics, from which a variety of technical ceramic materials have been developed, particularly for electrotechnical and electronic applications. Some silicate ceramics are also used for high-temperature applications in the processing of materials. Silicate ceramics are composed essentially of porcelain, steatite, cordierite, and mullite ($3Al_2O_3 \cdot 2SiO_2$).

Recent standards that classify and characterize technical silicate ceramics are listed in Tables 3.2-4 – 3.2-7.

Table 3.2-4 Properties of Alkali aluminium silicates according to DIN EN 60672 [2.6]

Mechanical properties	Symbol	Units	Designation					
			C 110	C 111	C 112	C 120	C 130	C 140
			Quartz porcelain, plastically formed	Quartz porcelain, pressed	Cristobalite porcelain	Alumina porcelain	Alumina porcelain, high-strength	Lithium porcelain
Density, minimum	ϱ	t/m^3	2	2.2	2.3	2.3	2.5	2.0
Bending strength, unglazed	σ_{ft}	MPa	50	40	80	90	140	50
Bending strength, glazed	σ_{fg}	MPa	60	–	100	110	160	60
Young's modulus	E	GPa	60	–	70	–	100	–
Electrical properties								
Breakdown strength	E_d	kV/mm	20	–	20	20	20	15
Withstand voltage	U	kV	30	–	30	30	30	20
Permittivity at 48–62 Hz	ε_r		6–7	–	5–6	6–7	6–7.5	5–7
Loss factor at 20 °C, 1 kHz	$\tan\delta_{pf}$	10^{-3}	25	–	25	25	30	10
Loss factor at 20 °C, 1 MHz	$\tan\delta_{1M}$	10^{-3}	12	–	12	12	15	10
Resistivity at 20 °C	ρ_{20}	Ω m	10^{11}	10^{10}	10^{11}	10^{11}	10^{11}	10^{11}
Resistivity at 600 °C	ρ_{600}	Ω m	10^2	10^2	10^2	10^2	10^2	10^2
Thermal properties								
Average coefficient of thermal expansion at 30–600 °C	α_{30-600}	10^{-6} K^{-1}	4–7	4–7	6–8	4–7	5–7	1–3
Specific heat capacity at 30–600 °C	$c_{p,30-600}$	J kg^{-1}K^{-1}	750–900	800–900	800–900	750–900	800–900	750–900
Thermal conductivity	λ_{30-100}	W m^{-1}K^{-1}	1–2.5	1.0–2.5	1.4–2.5	1.2–2.6	1.5–4.0	1.0–2.5
Thermal fatigue resistance		(Rated)	Good	Good	Good	Good	Good	Good

Table 3.2-5 Properties of magnesium silicate according to DIN EN 60672 [2.6]

			Designation					
			C 210	C 220	C 221	C 230	C 240	C 250
Mechanical properties	Symbol	Units	Steatite for low voltage	Steatite, standard	Steatite, low-loss	Steatite, porous	Forsterite, porous	Forsterite, dense
Open porosity		vol. %	0.5	0.0	0.0	35.0	30.0	0.0
Density, minimum	ϱ	t/m^3	2.3	2.6	2.7	1.8	1.9	2.8
Bending strength	σ_B	MPa	80	120	140	30	35	140
Young's modulus	E	GPa	60	80	110	–	–	–
Electrical properties								
Breakdown strength	E_d	kV/mm	–	15	20	–	–	20
Withstand voltage	U	kV	–	20	30	–	–	20
Permittivity at 48–62 Hz	ε_r		6	6	6	–	–	7
Loss factor at 20 °C, 1 kHz	$\tan\delta_{pf}$	10^{-3}	25	5	1.5	–	–	1.5
Loss factor at 20 °C, 1 MHz	$\tan\delta_{1M}$	10^{-3}	7	3	1.2	–	–	0.5
Resistivity at 20 °C	ρ_{20}	Ω m	10^{10}	10^{11}	10^{11}	–	–	10^{11}
Resistivity at 600 °C	ρ_{600}	Ω m	10^3	10^3	10^5	10^5	10^5	10^5
Thermal properties								
Average coefficient of thermal expansion at 30–600 °C	α_{30-600}	10^{-6} K^{-1}	6–8	7–9	7–9	8–10	8–10	9–11
Specific heat capacity at 30–100 °C	$c_{p,30-600}$	J kg^{-1}K^{-1}	800–900	800–900	800–900	800–900	800–900	800–900
Thermal conductivity	λ_{30-100}	W m^{-1}K^{-1}	1–2.5	2–3	2–3	1.5–2	1.4–2	3–4
Thermal fatigue resistance		(Rated)	Good	Good	Good	Good	Good	Good

Table 3.2-6 Properties of alkaline-earth aluminium silicates according to DIN EN 60672 [2.6]

			Designation			
			C 410	C 420	C 430	C 440
Mechanical properties	Symbol	Units	Cordierite, dense	Celsians, dense	Calcium-based, dense	Zirconium-based, dense
Open porosity		vol. %	0.5	0.5	0.5	0.5
Density, minimum	ϱ	t/m^3	2.1	2.7	2.3	2.5
Bending strength	σ_B	MPa	60	80	80	100
Young's modulus	E	GPa	–	–	80	130

Table 3.2-6 Properties of alkaline-earth aluminium silicates according to DIN EN 60672 [2.6], cont.

Electrical properties	Symbol	Units	Designation			
			C 410 Cordierite, dense	C 420 Celsians, dense	C 430 Calcium-based, dense	C 440 Zirconium-based, dense
Breakdown strength	E_d	kV/mm	10	20	15	15
Withstand voltage	U	kV	15	30	20	20
Permittivity at 48–62 Hz	ε_r		5	7	6–7	8–12
Loss factor at 20 °C, 1 kHz	$\tan \delta_{pf}$	10^{-3}	25	10	5	5
Loss factor at 20 °C, 1 MHz	$\tan \delta_{1M}$	10^{-3}	7	0.5	5.0	5
Resistivity at 20 °C	ρ_{20}	Ω m	10^{10}	10^{12}	10^{11}	10^{11}
Resistivity at 600 °C	ρ_{600}	Ω m	10^3	10^7	10^2	10^2
Thermal properties						
Average coefficient of thermal expansion at 30–600 °C	α_{30-600}	10^{-6} K^{-1}	2–4	3.5–6	–	–
Specific heat capacity at 30–100 °C	$c_{p,30-600}$	J kg^{-1} K^{-1}	800–1200	800–1000	700–850	550–650
Thermal conductivity	λ_{30-100}	W m^{-1} K^{-1}	1.2–2.5	1.5–2.5	1–2.5	5–8
Thermal fatigue resistance		(Rated)	Good	Good	Good	Good

Table 3.2-7 Properties of porous aluminium silicates and magnesium silicates according to DIN EN 60672 [2.6]

Mechanical properties	Symbol	Units	Designation				
			C 510 Aluminosilicate-based	C 511 Magnesia–aluminosilicate-based	C 512 Magnesia–aluminosilicate-based	C 520 Cordierite-based	C 530 Aluminosilicate-based
Open porosity		vol. %	30	20	40	20	30
Density, minimun	ϱ	t/m^3	1.9	1.9	1.8	1.9	2.1
Bending strength	σ_B	MPa	25	25	15	30	30
Young's modulus	E	GPa	–	–	–	40	–
Electrical properties							
Resistivity at 600 °C	ρ_{600}	Ω m	10^3	10^3	10^3	10^3	10^4
Thermal properties							
Average coefficient of thermal expansion at 30–600 °C	α_{30-600}	10^{-6} K^{-1}	3–6	4–6	3–6	2–4	4–6
Specific heat capacity at 30–100 °C	$c_{p,30-600}$	J kg^{-1} K^{-1}	750–850	750–850	750–900	750–900	800–900
Thermal conductivity	λ_{30-100}	W m^{-1} K^{-1}	1.2–1.7	1.3–1.8	1–1.5	1.3–1.8	1.4–2.0
Thermal fatigue resistance		Rated	Good	Good	Good	Good	Good

3.2.3 Refractory Ceramics

Traditional refractory ceramics are produced as bricks from a broad variety of materials for various applications, ranging from building bricks used at ambient and moderately elevated temperatures to high-temperature refractory grades with particularly high melting temperatures and refractory stability, such as magnesite, silicon carbide, stabilized zirconia, and chrome-magnesite. Table 3.2-8 gives a survey of fired refractory brick materials.

It should be noted that these refractory ceramics in brick form differ from materials classified as technical refractory ceramics in the amounts of material used and in the composition, including the impurity content, but they rely basically on the same oxide or non-oxide ceramic compounds. These compounds are used in technical ceramics and are treated in more detail below.

Table 3.2-8 Properties of fired refractory brick materials [2.4]

Brick (major chemical components)	Density ϱ (kg/m^3)	Melting temperature (°C)	Thermal conductivity κ (W/m K)
Alumina brick (6–65 wt% Al_2O_3)	1842	1650–2030	4.67
Building brick	1842	1600	0.72
Carbonbrick (99 wt% graphite)	1682	3500	3.6
Chromebrick (100 wt% Cr_2O_3)	2900–3100	1900	2.3
Crome–magnesite brick (52 wt% MgO, 23 wt% Cr_2O_3)	3100	3045	3.5
Fireclay brick (54 wt% SiO_2, 40 wt% Al_2O_3)	2146–2243	1740	0.3–1.0
Fired dolomite (55 wt% CaO, 37 wt% MgO)	2700	2000	–
High-alumina brick (90–99 wt% Al_2O_3)	2810–2970	1760–2030	3.12
Magnesite brick (95.5 wt% MgO)	2531–2900	2150	3.7–4.4
Mullite brick (71 wt% Al_2O_3)	2450	1810	7.1
Silica brick (95–99 wt% SiO_2)	1842	1765	1.5
Silicon carbide brick (80–90 wt% SiC)	2595	2305	20.5
Zircon (99 wt% $ZrSiO_4$)	3204	1700	2.6
Zirconia (stabilized) brick	3925	2650	2.0

3.2.4 Oxide Ceramics

Oxides are the most common constituents of all ceramics, traditional and technical. An extensive account is given in [2.8]. They are used for their physical as well as refractory properties. Table 3.2-9 presents a systematic listing of their composition, structure and properties. Some of the more important oxides are dealt with in more detail further below.

Table 3.2-9 Physical properties of oxides and oxide-based high-temperature refractories [2.4]

IUPAC name (synonyms and common trade names)	Theoretical chemical formula, [CASRN], relative molecular mass ($^{12}C = 12.000$)	Crystal system, lattice parameters, *Strukturbericht* symbol, Pearson symbol, space group, structure type, Z	Density (ϱ, kg m^{-3})	Electrical resistivity (ρ, $\mu\Omega$ cm)	Melting point (°C)	Thermal conductivity (κ, W m^{-1} K^{-1})	Specific heat capacity (c_p, J kg^{-1} K^{-1})	Coefficient linear thermal expansion (α, 10^{-6} K^{-1})
Aluminium sesquioxide (alumina, corundum, sapphire)	α-Al$_2$O$_3$ [1344-28-1] [1302-74-5] 101.961	Trigonal (rhombohedral) $a = 475.91$ pm $c = 1298.4$ pm D5$_1$, hR10, $R\bar{3}c$, corundum type ($Z = 2$)	3987	2×10^{23}	2054	35.6–39	795.5–880	7.1–8.3
Beryllium monoxide (beryllia)	BeO [1304-56-9] 25.011	Trigonal (hexagonal) $a = 270$ pm $c = 439$ pm B4, hP4, $P6_3mc$, wurtzite type ($Z = 2$)	3008–3030	10^{22}	2550–2565	245–250	996.5	7.5–9.7
Calcium monoxide (calcia, lime)	CaO [1305-78-8] 56.077	Cubic $a = 481.08$ pm B2, cP2, $Pm3m$, CsCl type ($Z = 1$)	3320	10^{14}	2927	8–16	753.1	3.88
Cerium dioxide (ceria, cerianite)	CeO$_2$ [1306-38-3] 172.114	Cubic $a = 541.1$ pm C1, cF12, $Fm3m$, fluorite type ($Z = 4$)	7650	10^{10}	2340	–	389	10.6
Chromium oxide (eskolaite)	Cr$_2$O$_3$ [1308-38-9] 151.990	Trigonal (rhombohedral) $a = 538$ pm $\alpha = 54°50'$ D5$_1$, hR10, $R\bar{3}c$, corundum type ($Z = 2$)	5220	1.3×10^9 (346 °C)	2330	–	921.1	10.9
Dysprosium oxide (dysprosia)	Dy$_2$O$_3$ [1308-87-8] 373.00	Cubic D5$_3$, cI80, $Ia\bar{3}$, Mn$_2$O$_3$ type ($Z = 16$)	8300	–	2408	–	–	7.74
Europium oxide (europia)	Eu$_2$O$_3$ [1308-96-9] 351.928	Cubic D5$_3$, cI80, $Ia\bar{3}$, Mn$_2$O$_3$ type ($Z = 16$)	7422	–	2350	–	–	7.02
Hafnium dioxide (hafnia)	HfO$_2$ [12055-23-1] 210.489	Monoclinic (1790°C) $a = 511.56$ pm $b = 517.22$ pm $c = 529.48$ pm C43, mP12, $P2_1c$, baddeleyite type ($Z = 4$)	9680	5×10^{15}	2900	1.14	121	5.85
Gadolinium oxide (gadolinia)	Gd$_2$O$_3$ [12064-62-9] 362.50	Cubic D5$_3$, cI80, $Ia\bar{3}$, Mn$_2$O$_3$ type ($Z = 16$)	7630	–	2420	–	276	10.44
Lanthanum dioxide (lanthana)	La$_2$O$_3$ [1312-81-8] 325.809	Trigonal (hexagonal) D5$_2$, hP5, $P\bar{3}m$1, lanthana type ($Z = 1$)	6510	10^{14} (550 °C)	2315	–	288.89	11.9

Table 3.2-9 Physical properties of oxides and oxide-based high-temperature refractories [2.4], cont.

Young's modulus (E, GPa)	Flexural strength (τ, MPa)	Compressive strength (α, MPa)	Vickers hardness HV (Mohs hardness HM)	Other physicochemical properties, corrosion resistance,[a] and uses	IUPAC name (synonyms and common trade names)
365–393	282	2549–3103	2100–3000 (HM 9)	White and translucent; hard material used as abrasive for grinding. Excellent electrical insulator and also wear-resistant. Insoluble in water, insoluble in strong mineral acids, readily soluble in strong solutions of alkali metal hydroxides, attacked by HF and NH_4HF_2. Owing to its corrosion resistance under an inert atmosphere in molten metals such as Mg, Ca, Sr, Ba, Mn, Sn, Pb, Ga, Bi, As, Sb, Hg, Mo, W, Co, Ni, Pd, Pt, and U, it is used for crucibles for these liquid metals. Alumina is readily attacked under an inert atmosphere by molten metals such as Li, Na, Be, Al, Si, Ti, Zr, Nb, Ta, and Cu. Maximum service temperature 1950 °C.	Aluminium sesquioxide (alumina, corundum, sapphire)
296.5–345	241–250	1551	1500 (HM 9)		Beryllium monoxide (beryllia)
–	–	–	560 (HM 4.5)	Forms white or grayish ceramics. It is readily absorbs CO_2 and water from air to form calcium carbonate and slaked lime. It reacts readily with water to give $Ca(OH)_2$. Volumetric expansion coefficient 0.225×10^{-9} K^{-1}. It exhibits outstanding corrosion resistance in the following liquid metals: Li and Na.	Calcium monoxide (calcia, lime)
181	–	589	(HM 6)	Pale yellow cubic crystals. Abrasive for polishing glass; used in interference filters and antireflection coatings. Insoluble in water, soluble in H_2SO_4 and HNO_3, but insoluble in dilute acids.	Cerium dioxide (ceria, Cerianite)
–	–	–	(HM > 8)		Chromium oxide (eskolaite)
–	–	–	–		Dysprosium oxide (dysprosia)
–	–	–	–		Europium oxide (europia)
57	–	–	780–1050		Hafnium dioxide (hafnia)
124	–	–	480		Gadolinium oxide (gadolinia)
–	–	–	–	Insoluble in water, soluble in dilute strong mineral acids.	Lanthanum dioxide (lanthana)

Table 3.2-9 Physical properties of oxides and oxide-based high-temperature refractories [2.4], cont.

IUPAC name (synonyms and common trade names)	Theoretical chemical formula, [CASRN], relative molecular mass ($^{12}C = 12.000$)	Crystal system, lattice parameters, *Strukturbericht* symbol, Pearson symbol, space group, structure type, Z	Density (ϱ, kg m^{-3})	Electrical resistivity (ρ, $\mu\Omega$ cm)	Melting point (°C)	Thermal conductivity (κ, W m^{-1} K^{-1})	Specific heat capacity (c_p, J kg^{-1} K^{-1})	Coefficient of linear thermal expansion (α, 10^{-6} K^{-1})
Magnesium monoxide (magnesia, periclase)	MgO [1309-48-4] 40.304	Cubic $a = 420$ pm B1, cF8, Fm3m, rock salt type (Z = 4)	3581	1.3×10^{15}	2852	50–75	962.3	11.52
Niobium pentoxide (columbite, niobia)	Nb$_2$O$_5$ [1313-96-8] 265.810	Numerous polytypes	4470	5.5×10^{12}	1520	–	502.41	–
Samarium oxide (samaria)	Sm$_2$O$_3$ [12060-58-1] 348.72	Cubic D5$_3$, c180, $Ia\bar{3}$, Mn$_2$O$_3$ type (Z = 16)	7620	–	2350	2.07	331	10.3
Silicon dioxide (silica, α-quartz)	α-SiO$_2$ [7631-86-9] [14808-60-7] 60.085	Trigonal (rhombohedral) $a = 491.27$ pm $c = 540.46$ pm C8, hP9, $R\bar{3}c$, α-quartz type (Z = 3)	2202–2650	10^{20}	1710	1.38	787	0.55
Tantalum pentoxide (tantalite, tantala)	Ta$_2$O$_5$ [1314-61-0] 441.893	Trigonal (rhombohedral) columbite type	8200	10^{12}	1882	–	301.5	–
Thorium dioxide (thoria, thorianite)	ThO$_2$ [1314-20-1] 264.037	Cubic $a = 559$ pm Cl, cF12, Fm3m, fluorite type (Z = 4)	9860	4×10^{19}	3390	14.19	272.14	9.54
Titanium dioxide (anatase)	TiO$_2$ [13463-67-7] [1317-70-0] 79.866	Tetragonal $a = 378.5$ pm $c = 951.4$ pm C5, tI12, $I4_1amd$, anatase type (Z = 4)	3900	–	700 °C (rutile)	–	–	–
Titanium dioxide (brookite)	TiO$_2$ [13463-67-7] 79.866	Orthorhombic $a = 916.6$ pm $b = 543.6$ pm $c = 513.5$ pm C21, oP24, Pbca, brookite type (Z = 8)	4140	–	1750	–	–	–

Table 3.2-9 Physical properties of oxides and oxide-based high-temperature refractories [2.4], cont.

Young's modulus (E, GPa)	Flexural strength (τ, MPa)	Compressive strength (α, MPa)	Vickers hardness HV (Mohs hardness HM)	Other physicochemical properties, corrosion resistance,[a] and uses	IUPAC name (synonyms and common trade names)
303.4	441	1300–1379	750 (HM 5.5–6)	Forms ceramics with a high reflection coefficient in the visible and near-UV region. Used in linings for steelmaking furnaces and in crucibles for fluoride melts. Very slowly soluble in pure water; but soluble in dilute strong mineral acids. It exhibits outstanding corrosion resistance in the following liquid metals: Mg, Li, and Na. It is readily attacked by molten metals such as Be, Si, Ti, Zr, Nb, and Ta. MgO reacts with water, CO_2, and dilute acids. Maximum service temperature 2400 °C. Transmittance of 80% and refractive index of 1.75 in the IR region from 7 to 300 μm.	Magnesium monoxide (magnesia, periclase)
–	–	–	1500	Dielectric used in film supercapacitors. Insoluble in water; soluble in HF and in hot concentrated H_2SO_4.	Niobium pentoxide (columbite, niobia)
183	–	–	438		Samarium oxide (samaria)
72.95	310	680–1380	550–1000 (HM 7)	Colorless amorphous (fused silica) or crystalline (quartz) material having a low thermal expansion coefficient and excellent optical transmittance in the far UV. Silica is insoluble in strong mineral acids and alkalis except HF, concentrated H_3PO_4, NH_4HF_2 and concentrated alkali metal hydroxides. Owing to its good corrosion resistance to liquid metals such as Si, Ge, Sn, Pb, Ga, In, Tl, Rb, Bi, and Cd, it is used in crucibles for melting these metals. Silica is readily attacked under an inert atmosphere by molten metals such as Li, Na, K, Mg, and Al. Quartz crystals are piezoelectric and pyroelectric. Maximum service temperature 1090 °C.	Silicon dioxide (silica, α-quartz)
–	–	–	–	Dielectric used in film supercapacitors. Tantalum oxide is a high-refractive index, low-absorption material usable for making optical coatings from the near-UV (350 nm) to the IR (8 μm). Insoluble in most chemicals except HF, HF–HNO_3 mixtures, oleum, fused alkali metal hydroxides (e.g. NaOH and KOH) and molten pyrosulfates.	Tantalum pentoxide (tantalite, tantala)
144.8	–	1475	945 (HM 6.5)	Corrosion-resistant container material for the following molten metals: Na, Hf, Ir, Ni, Mo, Mn, Th, U. Corroded by the following liquid metals: Be, Si, Ti, Zr, Nb, Bi. Radioactive.	Thorium dioxide (thoria, thorianite)
–	–	–	(HM 5.5–6)		Titanium dioxide (anatase)
–	–	–	(HM 5.5–6)		Titanium dioxide (brookite)

Table 3.2-9 Physical properties of oxides and oxide-based high-temperature refractories [2.4], cont.

IUPAC name (synonyms and common trade names)	Theoretical chemical formula, [CASRN], relative molecular mass ($^{12}C = 12.000$)	Crystal system, lattice parameters, Strukturbericht, symbol, Pearson symbol, space group, structure type, Z	Density (ϱ, kg m^{-3})	Electrical resistivity (ρ, $\mu\Omega$ cm)	Melting point (°C)	Thermal conductivity (κ, W m^{-1} K^{-1})	Specific heat capacity (c_p, J kg^{-1} K^{-1})	Coefficient of linear thermal expansion (α, 10^{-6} K^{-1})
Titanium dioxide (rutile, titania)	TiO$_2$ [13463-67-7] [1317-80-2] 79.866	Tetragonal $a = 459.37$ pm $c = 296.18$ pm C4, tP6, P4/mnm, rutile type ($Z = 2$)	4240	10^{19}	1855	10.4 (∥ c) 7.4 (⊥ c)	711	7.14
Uranium dioxide (uraninite)	UO$_2$ [1344-57-6] 270.028	Cubic $a = 546.82$ pm C1, cF12, Fm3m, fluorite type ($Z = 4$)	10 960	3.8×10^{10}	2880	10.04	234.31	11.2
Yttrium oxide (yttria)	Y$_2$O$_3$ [1314-36-9] 225.81	Trigonal (hexagonal) D5$_2$, hP5, $P\bar{3}m1$, lanthana type ($Z = 1$)	5030	–	2439	–	439.62	8.10
Zirconium dioxide (baddeleyite)	ZrO$_2$ [1314-23-4] [12036-23-6] 123.223	Monoclinic $a = 514.54$ pm $b = 520.75$ pm $c = 531.07$ pm $\alpha = 99.23°$ C43, mP12, P2$_1$c, baddeleyite type ($Z = 4$)	5850	–	2710	–	711	7.56
Zirconium dioxide PSZ (stabilized with MgO) (zirconia > 2300 °C)	ZrO$_2$ [1314-23-4] [64417-98-7] 123.223	Cubic C1, cF12, Fm3m, fluorite type ($Z = 4$)	5800–6045	–	2710	1.8	400	10.1
Zirconium dioxide TTZ (stabilized with Y$_2$O$_3$) (zirconia > 2300 °C)	ZrO$_2$ [1314-23-4] [64417-98-7] 123.223	Cubic C1, cF12, Fm3m, fluorite type ($Z = 4$)	6045	–	2710	–	–	–
Zirconium dioxide TZP (zirconia > 1170 °C)	ZrO$_2$ [1314-23-4] 123.223	Tetragonal C4, tP6, P4$_2$/mnm, rutile type ($Z = 2$)	5680–6050	7.7×10^7	2710	–	–	10–11
Zirconium dioxide (stabilized with 10–15% Y$_2$O$_3$) (zirconia > 2300 °C)	ZrO$_2$ [1314-23-4] [64417-98-7] 123.223	Cubic C1, cF12, Fm3m, fluorite type ($Z = 4$)	6045	–	2710	–	–	–

[a] Corrosion data in molten salts from [2.9].

Table 3.2-9 Physical properties of oxides and oxide-based high-temperature refractories [2.4], cont.

Young's modulus (E, GPa)	Flexural strength (τ, MPa)	Compressive strength (α, MPa)	Vickers hardness HV (Mohs hardness HM)	Other physicochemical properties, corrosion resistance,[a] and uses	IUPAC name (synonyms and common trade names)
248–282	340	800–940	(HM 7–7.5)	White, translucent, hard ceramic material. Readily soluble in HF and in concentrated H_2SO_4, and reacts rapidly with molten alkali hydroxides and fused alkali carbonates. Owing to its good corrosion resistance to liquid metals such as Ni and Mo, it is used in crucibles for melting these metals. Titania is readily attacked under an inert atmosphere by molten metals such as Be, Si, Ti, Zr, Nb, and Ta.	Titanium dioxide (rutile, titania)
145	–	–	600 (HM 6–7)	Used in nuclear power reactors in sintered nuclear-fuel elements containing either natural or enriched uranium.	Uranium dioxide (uraninite)
114.5	–	393	700	Yttria is a medium-refractive-index, low-absorption material usable for optical coating, in the near-UV (300 nm) to the IR (12 μm) regions. Hence used to protect Al and Ag mirrors. Used for crucibles containing molten lithium.	Yttrium oxide (yttria)
241	–	2068	(HM 6.5) 1200	Zirconia is highly corrosion-resistant to molten metals such as Bi, Hf, Ir, Pt, Fe, Ni, Mo, Pu, and V, while is strongly attacked by the following liquid metals: Be, Li, Na, K, Si, Ti, Zr, and Nb. Insoluble in water, but slowly soluble in HCl and HNO_3; soluble in boiling concentrated H_2SO_4 and alkali hydroxides and readily attacked by HF. Monoclinic (baddeleyite) below 1100 °C, tetragonal between 1100 and 2300 °C, cubic (fluorite type) above 2300 °C. Maximum service temperature 2400 °C.	Zirconium dioxide (baddeleyite)
200	690	1850	1600		Zirconium dioxide PSZ (stabilized with MgO) (zirconia > 2300 °C)
–	–	–	–		Zirconium dioxide TTZ (stabilized with Y_2O_3) (zirconia > 2300 °C)
200–210	> 800	> 2900	–		Zirconium dioxide TZP (zirconia > 1170 °C)
–	–	–	–		Zirconium dioxide (stabilized 10–15% Y_2O_3) (zirconia > 2300 °C)

3.2.4.1 Magnesium Oxide

Table 3.2-10 lists the properties of the magnesium oxide rich raw material used in the production of oxide ceramics. MgO occurs rarely, as the mineral periclase, but is abundant as magnesite ($MgCO_3$) and dolomite (($Mg,Ca)CO_3$).

Magnesium oxide is mainly processed into a high-purity single-phase ceramic material, in either porous or gas-tight form. Some properties of porous MgO are listed in Table 3.2-11. The properties of MgO ceramics that are of particular benefit in applications are their high electrical insulation capability and their high thermal conductivity. Typical examples of their application are for insulation in sheathed thermocouples and in resistive heating elements.

Table 3.2-10 Physical properties of MgO-rich raw materials used in the production of refractory ceramics [2.3]

Composition (wt%)	Sintered MgO clinker	Sintered doloma clinker	Fused MgO
MgO	96–99	39–40	97–98
Al_2O_3	0.05–0.25	0.3–0.8	0.1–0.2
Fe_2O_3	0.05–0.2	0.6–1.0	0.1–0.5
CaO	0.6–2.4	57.5–58.5	0.9–2.5
SiO_2	0.1–0.5	0.6–1.1	0.3–0.9
B_2O_3	0.005–0.6		0.004–0.01
Bulk density (g/cm^3)	3.4–3.45	3.15–3.25	3.5
Grain porosity (vol.%)	2–3	5.5–7	0.5–2

Table 3.2-11 Properties of MgO according to DIN EN 60672 [2.6]

	Symbol	Units	Designation C 820 Porous
Mechanical properties			
Open porosity		vol. %	30
Density, minimum	ϱ	Mg/m^3	2.5
Bending strength	σ_B	MPa	50
Young's modulus	E	GPa	90
Electrical properties			
Permittivity at 48–62 Hz	ε_r		10
Thermal properties			
Average coefficient of thermal expansion at 30–600 °C	α_{30-600}	10^{-6} K^{-1}	11–13
Specific heat capacity at 30–100 °C	$c_{p,30-600}$	J kg^{-1} K^{-1}	850–1050
Thermal conductivity	λ_{30-100}	W m^{-1} K^{-1}	6–10
Thermal fatigue resistance		(Rated)	Good

3.2.4.2 Alumina

The properties of commercial grades of alumina (corundum) are closely related to the microstructure. Pure α-Al_2O_3 is denser, harder, stiffer, and more refractory than most silicate ceramics so that increasing the proportion of the second phase in an alumina ceramic tends to decrease the density, Young's modulus, strength, hardness, and refractoriness. Sintered alumina is produced from high-purity powders, which densify to give single-phase ceramics with a uniform grain size. Table 3.2-12 gives typical examples of properties of high-density alumina, and Table 3.2-13 lists further properties of various alumina ceramics.

Alumina is widely applied both as an electronic ceramic material and in cases where its nonelectronic properties, such as fracture toughness, wear and high-temperature resistance, are required. Some typical applications are in insulators in electrotechnical equipment; substrates for electronic components; wear-resistant machine parts; refractory materials in the chemical industry, where resistance against vapors, melts, and slags is important; insulating materials in sheathed thermocouples and heating elements; medical implants; and high-temperature parts such as burner nozzles.

Table 3.2-12 Typical properties of high-density alumina [2.3]

	Al_2O_3 content (wt%)			
	> 99.9	> 99.7 [a]	> 99.7 [b]	99–99.7
Density (g/cm^3)	3.97–3.99	3.6–3.85	3.65–3.85	3.89–3.96
Hardness (GPa), HV 500 g	19.3	16.3	15–16	15–16
Fracture toughness K_{IC} at room temperature (MPa m$^{1/2}$)	2.8–4.5	–	–	5.6–6
Young's modulus (GPa)	366–410	300–380	300–380	330–400
Bending strength (MPa) at room temperature	550–600	160–300	245–412	550
Thermal expansion coefficient (10^{-6}/K) at 200–1200 °C	6.5–8.9	5.4–8.4	5.4–8.4	6.4–8.2
Thermal conductivity at room temperature (W/m K)	38.9	28–30	30	30.4
Firing temperature range (°C)	1600–2000	1750–1900	1750–1900	1700–1750

[a] "Recrystallized" without MgO.
[b] With MgO.

Table 3.2-13 Properties of Al_2O_3 according to DIN EN 60672 [2.6]

Mechanical properties	Symbol	Units	Designation							
			C 780	C 786	C 795	C 799	Al_2O_3	Al_2O_3	Al_2O_3	Al_2O_3
			Alumina 80–86%	Alumina 86–95%	Alumina 95–99%	Alumina >99%	Alumina <90%	Alumina 92–96%	Alumina 99%	Alumina >99%
Open porosity		vol. %	0.0	0.0	0.0	0.0	0	0	0	0
Density, min.	ϱ	Mg/m^3	3.2	3.4	3.5	3.7	> 3.2	3.4–3.8	3.5–3.9	3.75–3.98
Bending strength	σ_B	MPa	200	250	280	300	> 200	230–400	280–400	300–580
Young's modulus	E	GPa	200	220	280	300	> 200	220–340	220–350	300–380
Vickers hardness	HV	100	–	–	–	–	12–15	12–15	12–20	17–23
Fracture toughness	K_{IC}	MPa $\sqrt{m}$	–	–	–	–	3.5–4.5	4–4.2	4–4.2	4–5.5

Table 3.2-13 Properties of Al_2O_3 according to DIN EN 60672 [2.6], cont.

Thermal properties Electrical properties	Symbol	Units	Designation C 780 Alumina 80–86%	C 786 Alumina 86–95%	C 795 Alumina 95–99%	C 799 Alumina >99%	Al_2O_3 Alumina <90%	Al_2O_3 Alumina 92–96%	Al_2O_3 Alumina 99%	Al_2O_3 Alumina >99%
Breakdown strength	E_d	kV/mm	10	15	15	17	10	15–25	15	17
Withstand voltage	U	kV	15	18	18	20	15	18	18	20
Permittivity at 48–62 Hz	ε_r	–	8	9	9	9	9	9–10	9	9
Loss factor at 20 °C, 1 kHz	$\tan \delta_{pf}$	10^{-3}	1.0	0.5	0.5	0.2	0.5–1.0	0.3–0.5	0.2–0.5	0.2–0.5
Loss factor at 20 °C, 1 MHz	$\tan \delta_{1M}$	10^{-3}	1.5	1	1	1	1	1	1	1
Resistivity at 20 °C	ρ_{20}	$\Omega\,m$	10^{12}	10^{12}	10^{12}	10^{12}	10^{12}–10^{13}	10^{12}–10^{14}	10^{12}–10^{15}	10^{12}–10^{15}
Resistivity at 600 °C	ρ_{600}	$\Omega\,m$	10^5	10^6	10^6	10^6	10^6	10^6	10^6	10^6
Ave. coeff. of thermal expansion at 30–600 °C	α_{30-600}	$10^{-6}\,K^{-1}$	6–8	6–8	6–8	7–8	6–8	6–8	6–8	7–8
Specific heat at 30–100 °C	$c_{p,30-100}$	$J\,kg^{-1}K^{-1}$	850–1050	850–1050	850–1050	850–1050	850–1050	850–1050	850–1050	850–1050
Thermal conductivity	λ_{30-100}	$W\,m^{-1}K^{-1}$	10–16	14–24	16–28	19–30	10–16	14–25	16–28	19–30
Thermal fatigue resistance		(Rated)	Good	Good	Good	Good	Good	Good	Good	Good
Typ. max. application temperature	T	°C	–	–	–	–	1400–1500	1400–1500	1400–1500	1400–1700

3.2.4.3 Al–O–N Ceramics

A class of variants of the ceramics related to alumina is based on the ternary Al–O–N system, which forms 13 different compounds, some of which are polytypes of aluminium nitride based on the wurtzite structure. Table 3.2-14 lists some typical properties compiled from various publications. Al–O–N ceramics have optical and dielectric properties which make them suitable, for use, for example, in windows for elctromagnetic radiation.

Table 3.2-14 Selected properties of Al–O–N ceramics [2.3]

Property	Value
Refractive index at $\lambda = 0.55\,\mu m$	1.77–1.88
IR cutoff	5.12 μm
UV cutoff	0.27 μm
Dielectric constant at 20 °C, 100 Hz	8.5
Dielectric constant at 500 °C, 100 Hz	14.0
Loss tangent at 20 °C, 100 Hz	0.002
Loss tangent at 500 °C, 100 Hz	1.0
Thermal conductivity at 20 °C	10.89 $\mathrm{W\,m^{-1}\,K^{-1}}$
Thermal expansion coefficient at 25–1000 °C	$7.6 \times 10^6\,\mathrm{K^{-1}}$
Flexural strength at 20 °C	306 MPa
Flexural strength at 1000 °C	267 MPa
Fracture toughness	2.0–2.9 $\mathrm{MPa\,m^{1/2}}$

3.2.4.4 Beryllium Oxide

Beryllium oxide (BeO, beryllia) is the only material apart from diamond which combines high thermal-shock resistance, high electrical resistivity, and high thermal conductivity at a similar level. Hence its major application is in heat sinks for electronic components. BeO is highly soluble in water, but dissolves slowly in concentrated acids and alkalis. It is highly toxic. It is highly corrosion-resistant in several liquid metals, such as Li, Na, Al, Ga, Pb, Ni, and Ir. Its maximum service temperature is 2400 °C. Some properties are listed in Table 3.2-15.

Table 3.2-15 Properties of BeO according to DIN EN 60672 [2.6]

			Designation C 810
Mechanical properties	Symbol	Units	Beryllium oxide, dense
Open porosity		vol. %	0.0
Density, minimum	ϱ	Mg/m^3	2.8
Bending strength	σ_B	MPa	150
Young's modulus	E	GPa	300

Table 3.2-15 Properties of BeO according to DIN EN 60672 [2.6], cont.

			Designation C 810
Electrical properties	Symbol	Units	Beryllium oxide, dense
Breakdown strength	E_d	kV/mm	13
Withstand voltage	U	kV	20
Permittivity at 48–62 Hz	ε_r		7
Loss factor at 20 °C, 1 kHz	$\tan \delta_{pf}$	10^{-3}	1
Loss factor at 20 °C, 1 MHz	$\tan \delta_{1M}$	10^{-3}	1
Resistivity at 20 °C	ρ_{20}	Ω m	10^{12}
Resistivity at 600 °C	ρ_{600}	Ω m	10^{7}
Thermal properties			
Average coefficient of thermal expansion at 30–600 °C	α_{30-600}	10^{-6} K^{-1}	7–8.5
Specific heat capacity at 30–100 °C	$c_{p,30-100}$	J kg^{-1} K^{-1}	1000–1250
Thermal conductivity	λ_{30-100}	W m^{-1} K^{-1}	150–220
Thermal fatigue resistance		(Rated)	Good

3.2.4.5 Zirconium Dioxide

Zirconium dioxide (ZrO$_2$), commonly called zirconium oxide, forms three phases, with a monoclinic, a tetragonal, and a cubic crystal structure. Dense parts may be obtained by sintering of the cubic or tetragonal phase only. In order to stabilize the cubic phase, "stabilizers" such as MgO, CaO, Y$_2$O$_3$, and CeO$_2$ are added.

An important class of zirconium dioxide ceramics is known as partially stabilized zirconia (PSZ), and Table 3.2-16 gives typical property data for several different grades. Further properties of PSZ are listed in Table 3.2-17. It should be noted that, as with most ceramic materials, the K_{1C} data depend on the test method applied, and all the other mechanical properties are strongly affected by the microstructure (grain size, stabilizer content, etc.) and further variables such as temperature and atmospheric conditions.

Table 3.2-16 Physical properties reported for commercial grades of partially stabilized zirconia (PSZ) [2.3]

	Mg-PSZ	Ca-PSZ	Y-PSZ	Ca/Mg-PSZ
wt% stabilizer	2.5–3.5	3–4.5	5–12.5	3
Hardness (GPa)	14.4[a]	17.1[b]	13.6[c]	15
Fracture toughness K_{IC} at room temperature (MPa m$^{1/2}$)	7–15	6–9	6	4.6
Young's modulus (GPa)	200[a]	200–217	210–238	–
Bending strength (MPa) at room temperature	430–720	400–690	650–1400	350
Thermal expansion coefficient (10^{-6}/K) at 1000 °C	9.2[a]	9.2[b]	10.2[c]	–
Thermal conductivity at room temperature (W m^{-1} K^{-1})	1–2	1–2	1–2	1–2

[a] 2.8% MgO.
[b] 4% CaO.
[c] 5% Y$_2$O$_3$.

Table 3.2-17 Properties of PSZ according to DIN EN 60672 [2.6]

			Designation PSZ
Mechanical properties	**Symbol**	**Units**	**Partially stabilized ZrO_2**
Open porosity		vol. %	0
Density, minimum	ϱ	Mg/m^3	5–6
Bending strength	σ_B	MPa	500–1000
Young's modulus	E	GPa	200–210
Vickers hardness	HV	100	11–12.5
Fracture toughness	K_{IC}	MPa $\sqrt{m}$	5.8–10.5
Electrical properties			
Permittivity at 48–62 Hz	ε_r		22
Resistivity at 20 °C	ρ_{20}	Ω m	$10^8 – 10^{13}$
Resistivity at 600 °C	ρ_{600}	Ω m	$10^3 – 10^6$
Thermal properties	**Symbol**	**Units**	**Partially stabilized ZrO_2**
Average coefficient of thermal expansion at 30–600 °C	$\alpha_{30-1000}$	10^{-6} K^{-1}	10–12.5
Specific heat capacity at 30–100 °C	$c_{p,30-1000}$	J kg^{-1} K^{-1}	400–550
Thermal conductivity	λ_{30-100}	W m^{-1} K^{-1}	1.5–3
Thermal fatigue resistance		(Rated)	Good
Typical maximum application temperature	T	°C	900–1600

3.2.4.6 Titanium Dioxide, Titanates, etc.

Titanium dioxide (TiO_2) is widely used in powder form as a pigment and filler material and in optical and catalytic applications. Technical ceramics made from titanium dioxide or titanates have the characteristic feature that their permittivity and its temperature coefficient can be adjusted over a wide range. Furthermore, they have a low loss factor. Some properties are listed in Table 3.2-18.

Table 3.2-18 Properties of titanium dioxide, titanates etc. according to DIN EN 60672 [2.6]

Mechanical properties	Symbol	Units	Designation C 310 Titanium dioxide chief constituent	C 320 Magnesium titanate	C 330 Titanium dioxide with other oxides	C 331 Titanium dioxide with other oxides	C 340 Bismuth titanate, basic	C 350 Perovskites, middle ε_r
Open porosity		vol. %	0.0	0.0	0.0	0.0	0.0	0.0
Density, minimum	ϱ	Mg/m^3	3.5	3.1	4.0	4.5	3.0	4.0
Bending strength	σ_B	MPa	70	70	80	80	70	50
Breakdown strength	E_d	kV/mm	8	8	10	10	6	2
Withstand voltage	U	kV	15	15	15	15	8	2
Permittivity at 48–62 Hz	ε_r		40–100	12–40	25–50	30–70	100–700	350–3000
Loss factor at 20 °C, 1 kHz	$\tan \delta_{1k}$	10^{-3}	6.5	2	20.0	7	–	–
Loss factor at 20 °C, 1 MHz	$\tan \delta_{1M}$	10^{-3}	2	1.5	0.8	1	5	35.0
Resistivity at 20 °C	ρ_{20}	Ω m	10^{10}	10^9	10^9	10^9	10^9	10^8
Thermal properties								
Average coefficient of thermal expansion at 30–600 °C	α_{30-600}	10^{-6} K^{-1}	6–8	6–10	–	–	–	–
Specific heat capacity at 30–100 °C	$c_{p,30-100}$	J kg^{-1}K^{-1}	700–800	900–1000	–	–	–	–
Thermal conductivity	λ_{30-100}	W m^{-1}K^{-1}	3–4	3.5–4	–	–	–	–

3.2.5 Non-Oxide Ceramics

The non-oxide ceramics comprise essentially borides, carbides, nitrides, and silicides. Like oxide ceramics they have two kinds of uses, which frequently overlap: application of their physical properties and of their refractory high-temperature properties. Extensive accounts can be found in [2.1–3, 8, 10].

3.2.5.1 Non-Oxide High-Temperature Ceramics

The two dominating design variables to be considered for high-temperature materials are hardness and thermal conductivity as a function of temperature. Figure 3.2-2 gives a survey. It is obvious that borides and carbides are superior to most oxide ceramics. Depending upon the microstructure, SiC has a higher hardness than that of β-Si_3N_4, but it decreases somewhat more rapidly, with increasing temperature. Both Si_3N_4 and SiC ceramics possess a high thermal conductivity and thus excellent thermal-shock resistance.

3.2.5.2 Borides

Borides and boride-based high-temperature refractories are treated extensively in [2.2, 3, 10]. Some properties are listed in Table 3.2-19.

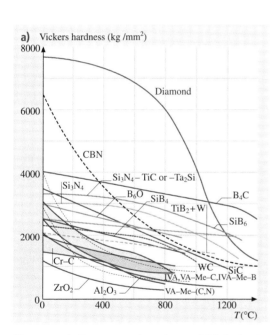
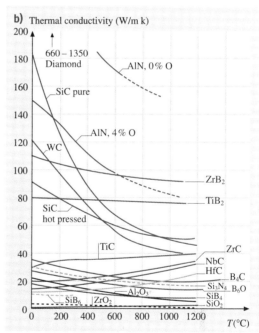

Fig. 3.2-2a,b Temperature dependence of (**a**) the hardness and (**b**) the thermal conductivity of ceramic materials in comparison with diamond and cubic boron nitride (CBN) [2.3]

Table 3.2-19 Physical properties of borides and boride-based high-temperature refractories [2.4]

IUPAC name	Theoretical chemical formula, [CASRN], relative molecular mass ($^{12}C = 12.000$)	Crystal system, lattice parameters, Strukturbericht symbol, Pearson symbol, space group, structure type, Z	Density (ϱ, kg m^{-3})	Electrical resistivity (ρ, $\mu\Omega$ cm)	Melting point (°C)	Thermal conductivity (κ, W m^{-1} K^{-1})	Specific heat capacity (c_p, J kg^{-1} K^{-1})	Coefficient of linear thermal expansion (α, 10^{-6} K^{-1})
Aluminium diboride	AlB$_2$ [12041-50-8] 48.604	Hexagonal $a = 300.50$ pm $c = 325.30$ pm C32, $hP3$, $P/6mmm$, AlB$_2$ type ($Z = 1$)	3190	–	1654	–	897.87	–
Aluminium dodecaboride	AlB$_{12}$ [12041-54-2] 156.714	Tetragonal $a = 1016$ pm $c = 1428$ pm	2580	–	2421	–	954.48	–
Beryllium boride	Be$_4$B [12536-52-6] 46.589	–	–	–	1160	–	–	–
Beryllium diboride	BeB$_2$ [12228-40-9] 30.634	Hexagonal $a = 979$ pm $c = 955$ pm	2420	10 000	1970	–	–	–
Beryllium hemiboride	Be$_2$B [12536-51-5]	Cubic $a = 467.00$ pm C1, $cF12$, $Fm3m$, CaF$_2$ type ($Z = 4$)	1890	1000	1520	–	–	–
Beryllium hexaboride	BeB$_6$ [12429-94-6]	Tetragonal $a = 1016$ pm $c = 1428$ pm	2330	10^{13}	2070	–	–	–
Beryllium monoboride	BeB [12228-40-9]	–	–	–	1970	–	–	–
Boron	β-B [7440-42-8] 10.811	Trigonal (rhombohedral) $a = 1017$ pm $\alpha = 65°$ 12' $hR105$, $R3m$, β-B type	2460	18 000	2190	–	–	–
Chromium boride	Cr$_5$B$_3$ [12007-38-4] 292.414	Orthorhombic $a = 302.6$ pm $b = 1811.5$ pm $c = 295.4$ pm D8$_1$, $tI32$, $I4/mcm$, Cr$_5$B$_3$ type ($Z = 4$)	6100	–	1900	15.8	–	13.7
Chromium diboride	CrB$_2$ [12007-16-8] 73.618	Hexagonal $a = 292.9$ pm $c = 306.6$ pm C32, $hP3$, $P6/mmm$, AlB$_2$ type ($Z = 1$)	5160–5200	21	1850–2100	20–32	712	6.2–7.5
Chromium monoboride	CrB [12006-79-0] 62.807	Tetragonal $a = 294.00$ pm $c = 1572.00$ pm B$_f$, $oC8$, $Cmcm$, CrB type ($Z = 4$)	6200	64.0	2000	20.1	–	12.3

Table 3.2-19 Physical properties of borides and boride-based high-temperature refractories [2.4], cont.

Young's modulus (E, GPa)	Flexural strength (τ, MPa)	Compressive strength (α, MPa)	Vickers hardness HV (Mohs hardness HM)	Other physicochemical properties, corrosion resistance,[a] and uses	IUPAC name
–	–	–	2500	Phase transition to AlB_{12} at 920 °C. Soluble in dilute HCl. Nuclear shielding material.	Aluminium diboride
–	–	–	–	Soluble in hot HNO_3, insoluble in other acids and alkalis. Neutron-shielding material.	Aluminium dodecaboride
–	–	–	–		Beryllium boride
–	–	–	–		Beryllium diboride
–	–	–	870		Beryllium hemiboride
–	–	–	–		Beryllium hexaboride
–	–	–	–		Beryllium monoboride
320	–	–	2055 (11)	Brown or dark powder, unreactive to oxygen, water, acids and alkalis. $\Delta H_{vap} = 480\,\text{kJ mol}^{-1}$.	Boron
–	–	–	–		Chromium boride
211	607	1300	1800	Strongly corroded by molten metals such as Mg, Al, Na, Si, V, Cr, Mn, Fe, and Ni. It is corrosion-resistant to the following liquid metals: Cu, Zn, Sn, Rb, and Bi.	Chromium diboride
–	–	–	–		Chromium monoboride

Table 3.2-19 Physical properties of borides and boride-based high-temperature refractories [2.4], cont.

IUPAC name	Theoretical chemical formula, [CASRN], relative molecular mass ($^{12}C = 12.000$)	Crystal system, lattice parameters, *Strukturbericht* symbol, Pearson symbol, space group, structure type, Z	Density (ϱ, kg m^{-3})	Electrical resistivity (ρ, $\mu\Omega$ cm)	Melting point (°C)	Thermal conductivity (κ, W m^{-1} K^{-1})	Specific heat capacity (c_p, J kg^{-1} K^{-1})	Coefficient of linear thermal expansion (α, 10^{-6} K^{-1})
Hafnium diboride	HfB$_2$ [12007-23-7] 200.112	Hexagonal $a = 314.20$ pm $c = 347.60$ pm C32, $hP3$, $P6/mmm$, AlB$_2$ type ($Z = 1$)	11 190	8.8–11	3250–3380	51.6	247.11	6.3–7.6
Lanthanum hexaboride	LaB$_6$ [12008-21-8] 203.772	Cubic $a = 415.7$ pm D2$_1$, $cP7$, $Pm3m$, CaB$_6$ type ($Z = 1$)	4760	17.4	2715	47.7	–	6.4
Molybdenum boride	Mo$_2$B$_5$ [12007-97-5] 245.935	Trigonal $a = 301.2$ pm $c = 2093.7$ pm D8$_1$, $hR7$, $R\bar{3}m$, Mo$_2$B$_5$ type ($Z = 1$)	7480	22–55	1600	50	–	8.6
Molybdenum diboride	MoB$_2$ 117.59	Hexagonal $a = 305.00$ pm $c = 311.30$ pm C32, $hP3$, $P6/mmm$, AlB$_2$ type ($Z = 1$)	7780	45	2100	–	527	7.7
Molybdenum hemiboride	Mo$_2$B [12006-99-4] 202.691	Tetragonal $a = 554.3$ pm $c = 473.5$ pm C16, $tI2$, $I4/mcm$, CuAl$_2$ type ($Z = 4$)	9260	40	2280	–	377	5
Molybdenum monoboride	MoB 106.77	Tetragonal $a = 311.0$ $c = 169.5$ B$_g$, $tI4$, $I4_1/amd$, MoB type ($Z = 2$)	8770	α-MoB 45, β-MoB 24	2180	–	368	–
Niobium diboride	ϵ-NbB$_2$ [12007-29-3] 114.528	Hexagonal $a = 308.90$ pm $c = 330.03$ pm C32, $hP3$, $P6/mmm$, AlB$_2$ type ($Z = 1$)	6970	26–65	2900	17–23.5	418	8.0–8.6
Niobium monoboride	δ-NbB [12045-19-1] 103.717	Orthorhombic $a = 329.8$ pm $b = 316.6$ pm $c = 87.23$ pm B$_f$, $oC8$, $Cmcm$, CrB type ($Z = 4$)	7570	40–64.5	2270–2917	15.6	–	12.9
Silicon hexaboride	SiB$_6$ [12008-29-6]	Trigonal (rhombohedral)	2430	200 000	1950	–	–	–
Silicon tetraboride	SiB$_4$ [12007-81-7] 71.330	–	2400	–	– (dec.)	–	–	–
Tantalum diboride	TaB$_2$ [12077-35-1] 202.570	Hexagonal $a = 309.80$ pm $c = 324.10$ pm C$_{32}$, $hP3$, $P6/mmm$, AlB$_2$ type ($Z = 1$)	12 540	33	3037–3200	10.9–16.0	237.55	8.2–8.8

Table 3.2-19 Physical properties of borides and boride-based high-temperature refractories [2.4], cont.

Young's modulus (E, GPa)	Flexural strength (τ, MPa)	Compressive strength (α, MPa)	Vickers hardness HV (Mohs hardness HM)	Other physicochemical properties, corrosion resistance,[a] and uses	IUPAC name
500	350	–	2900	Gray crystals, attacked by HF, otherwise highly resistant.	Hafnium diboride
479	126	–	–	Wear-resistant, semiconducting, thermoionic-conductor films.	Lanthanum hexaboride
672	345	–	–	Corroded by the following molten metals: Al, Mg, V, Cr, Mn, Fe, Ni, Cu, Nb, Mo, and Ta. It is corrosion resistant to molten Cd, Sn, Bi, and Rb.	Molybdenum boride
–	–	–	1280		Molybdenum diboride
–	–	–	– (HM 8–9)	Corrosion-resistant films.	Molybdenum hemiboride
–	–	–	1570		Molybdenum monoboride
637	–	–	3130 (HM > 8)	Corrosion-resistant to molten Ta; corroded by molten Re.	Niobium diboride
–	–	–	–	Wear-resistant, semiconducting films; neutron-absorbing layers on nuclear fuel pellets.	Niobium monoboride
–	–	–	–		Silicon hexaboride
–	–	–	–		Silicon tetraboride
257	–	–	(HM > 8)	Gray metallic powder. Severe oxidation in air above 800 °C. Corroded by the following molten metals: Nb, Mo, Ta, and Re.	Tantalum diboride

Table 3.2-19 Physical properties of borides and boride-based high-temperature refractories [2.4], cont.

IUPAC name	Theoretical chemical formula, [CASRN], relative molecular mass ($^{12}C = 12.000$)	Crystal system, lattice parameters, *Strukturbericht* symbol, Pearson symbol, space group, structure type, Z	Density (ϱ, kg m^{-3})	Electrical resistivity (ρ, $\mu\Omega$ cm)	Melting point (°C)	Thermal conductivity (κ, W m^{-1} K^{-1})	Specific heat capacity (c_p, J kg^{-1} K^{-1})	Coefficient of linear thermal expansion (α, 10^{-6} K^{-1})
Tantalum monoboride	TaB [12007-07-7] 191.759	Orthorhombic $a = 327.6$ pm $b = 866.9$ pm $c = 315.7$ pm B_f, $oC8$, $Cmcm$, CrB type ($Z = 4$)	14 190	100	2340–3090	–	246.85	–
Thorium hexaboride	ThB$_6$ [12229-63-9] 296.904	Cubic $a = 411.2$ pm $D2_1$, $cP7$, $Pm3m$, CaB$_6$ type ($Z = 1$)	6800	–	2149	44.8	–	7.8
Thorium tetraboride	ThB$_4$ [12007-83-9] 275.53	Tetragonal $a = 725.6$ pm $c = 411.3$ pm $D1_e$, $tP20$, $P4/mbm$, ThB$_4$ type ($Z = 4$)	8450	–	2500	25	510	7.9
Titanium diboride	TiB$_2$ [12045-63-5] 69.489	Hexagonal $a = 302.8$ pm $c = 322.8$ pm C32, $hP3$, $P6/mmm$, AlB$_2$ type ($Z = 1$)	4520	16–28.4	2980–3225	64.4	637.22	7.6–8.6
Tungsten hemiboride	W$_2$B [12007-10-2] 378.491	Tetragonal $a = 556.4$ pm $c = 474.0$ pm C16, $tI12$, $I4/mcm$, CuAl$_2$ type ($Z = 4$)	16 720	–	2670	–	168	6.7
Tungsten monoboride	WB [12007-09-9] 194.651	Tetragonal $a = 311.5$ pm $c = 1692$ pm	15 200 16 000	4.1	2660	–	–	6.9
Uranium diboride	UB$_2$ [12007-36-2] 259.651	Hexagonal $a = 313.10$ pm $c = 398.70$ pm C32, $hP3$, $P6/mmm$, AlB$_2$ type	12 710	–	2385	51.9	–	9
Uranium dodecaboride	UB$_{12}$ 367.91	Cubic $a = 747.3$ pm $D2_f$, $cF52$, $Fm3m$, UB$_{12}$ type ($Z = 4$)	5820	–	1500	–	–	4.6
Uranium tetraboride	UB$_4$ [12007-84-0] 281.273	Tetragonal $a = 707.5$ pm $c = 397.9$ pm $D1_e$, $tP20$, $P4/mbm$, ThB$_4$ type ($Z = 4$)	5350	–	2495	4.0	–	7.0
Vanadium diboride	VB$_2$ [12007-37-3] 72.564	Hexagonal $a = 299.8$ pm $c = 305.7$ pm C32, $hP3$, $P6/mmm$, AlB$_2$ type ($Z = 1$)	5070	23	2450–2747	42.3	647.43	7.6–8.3

Table 3.2-19 Physical properties of borides and boride-based high-temperature refractories [2.4], cont.

Young's modulus (E, GPa)	Flexural strength (τ, MPa)	Compressive strength (α, MPa)	Vickers hardness HV (Mohs hardness HM)	Other physicochemical properties, corrosion resistance,[a] and uses	IUPAC name
–	–	–	2200 (HM > 8)	Severe oxidation above 1100–1400 °C in air.	Tantalum monoboride
–	–	–	–		Thorium hexaboride
148	137	–	–		Thorium tetraboride
372–551	240	669	3370 (HM > 9)	Gray crystals, superconducting at 1.26 K. High-temperature electrical conductor, used in the form of a cermet as a crucible material for handling molten metals such as Al, Zn, Cd, Bi, Sn, and Rb. It is strongly corroded by liquid metals such as Tl, Zr, V, Nb, Ta, Cr, Mn, Fe, Co, Ni, and Cu. Begins to be oxidized in air above 1100–1400 °C. Corrosion-resistant in hot concentrated brines. Maximum operating temperature 1000 °C (reducing environment) and 800 °C (oxidizing environment).	Titanium diboride
–	–	–	2420 (HM 9)	Black powder.	Tungsten hemiboride
–	–	–	(HM 9)	Black powder.	Tungsten monoboride
–	–	–	1390		Uranium diboride
–	–	–	–		Uranium dodecaboride
440	413	–	2500		Uranium tetraboride
268	–	–	(HM 8–9)	Wear-resistant, semiconducting films.	Vanadium diboride

Table 3.2-19 Physical properties of borides and boride-based high-temperature refractories [2.4], cont.

IUPAC name	Theoretical chemical formula, [CASRN], relative molecular mass ($^{12}C = 12.000$)	Crystal system, lattice parameters, *Strukturbericht* symbol, Pearson symbol, space group, structure type, Z	Density (ϱ, kg m^{-3})	Electrical resistivity (ρ, $\mu\Omega$ cm)	Melting point (°C)	Thermal conductivity (κ, W m^{-1} K^{-1})	Specific heat capacity (c_p, J kg^{-1} K^{-1})	Coefficient of linear thermal expansion (α, 10^{-6} K^{-1})
Zirconium diboride	ZrB$_2$ [12045-64-6] 112.846	Hexagonal $a = 316.9$ pm $c = 353.0$ pm C32, hP3, $P6/mmm$, AlB$_2$ type (Z = 1)	6085	9.2	3060–3245	57.9	392.54	5.5–8.3
Zirconium dodecaboride	ZrB$_{12}$ 283.217	Cubic $a = 740.8$ pm D2$_f$, cF52, $Fm3m$, UB$_{12}$ type (Z = 4)	3630	60–80	2680	–	523	–

3.2.5.3 Carbides

Carbides are treated extensively in [2.1–3]. Some properties of carbides and carbide-based high-temperature refractories are listed in Table 3.2-20 and Table 3.2-21. It should be noted that carbides also play a major role as hardening constituents in all carbon-containing steels (Sect. 3.1.5).

Table 3.2-20 Physical properties of carbides and carbide-based high-temperature refractories [2.4]

IUPAC name (synonyms and common trade names)	Theoretical chemical formula, [CASRN], relative molecular mass ($^{12}C = 12.000$)	Crystal system, lattice parameters, *Strukturbericht* symbol, Pearson symbol, space group, structure type, Z	Density (ϱ, kg m^{-3})	Electrical resistivity (ρ, $\mu\Omega$ cm)	Melting point (°C)	Thermal conductivity (κ, W m^{-1} K^{-1})	Specific heat capacity (c_p, J kg^{-1} K^{-1})	Coefficient of linear thermal expansion (α, 10^{-6} K^{-1})
Aluminium carbides	Al$_4$C$_3$ [1299-86-1] 143.959	Numerous polytypes	2360	–	2798	n.a	n.a	n.a
Beryllium carbide	Be$_2$C [506-66-1] 30.035	Cubic $a = 433$ pm C1, cF12, $Fm3m$, CaF$_2$ type (Z = 4)	1900	–	2100	21.0	1397	10.5
Boron carbide (Norbide®)	B$_4$C [12069-32-8] 55.255	Hexagonal $a = 560$ pm $c = 1212$ pm DI$_g$, hR15, $R\bar{3}m$, B$_4$C type	2512	4500	2350–2427	27	1854	2.63–5.6

Table 3.2-19 Physical properties of borides and boride-based high-temperature refractories [2.4], cont.

Young's modulus (E, GPa)	Flexural strength (τ, MPa)	Compressive strength (α, MPa)	Vickers hardness HV (Mohs hardness HM)	Other physicochemical properties, corrosion resistance,[a] and uses	IUPAC name
343–506	305	–	1900–3400 (HM 8)	Gray metallic crystals, excellent thermal-shock resistance, greatest oxidation inertness of all refractory hard metals. Hot-pressed material is used in crucibles for handling molten metals such as Zn, Mg, Fe, Cu, Zn, Cd, Sn, Pb, Rb, Bi, Cr, brass, carbon steel, and cast iron, and also molten cryolite, yttria, zirconia, and alumina. It is readily corroded by liquid metals such as Si, Cr, Mn, Co, Ni, Nb, Mo, and Ta, and attacked by molten salts such as Na_2O, alkali carbonates, and NaOH. Severe oxidation in air occurs above 1100–1400 °C. Stable above 2000 °C under inert or reducing atmosphere.	Zirconium diboride
–	–	–	–		Zirconium dodecaboride

rosion data in molten salts from [2.9].

Table 3.2-20 Physical properties of carbides and carbide-based high-temperature refractories [2.4], cont.

Young's modulus (E, GPa)	Flexural strength (τ, MPa)	Compressive strength (α, MPa)	Vickers hardness HV (Mohs hardness HM)	Other physicochemical properties, corrosion resistance,[a] and uses	IUPAC name (synonyms and common trade names)
n.a	n.a	n.a	n.a	Decomposed in water with evolution of CH_4.	Aluminium carbides
314.4	–	723	2410 HK	Brick-red or yellowish-red octahedra. Used in nuclear-reactor cores.	Beryllium carbide
440–470	–	2900	3200–3500 HK (HM 9)	Hard, black, shiny crystals, the fourth hardest material known after diamond, cubic boron nitride, and boron oxide. It does not burn in an O_2 flame if the temperature is maintained below 983 °C. Maximum operating temperature 2000 °C (inert or reducing environment) or 600 °C (oxidizing environment). It is not attacked by hot HF or chromic acid. Used as abrasive, and in crucibles for molten salts, except molten alkalimetal hydroxides. In the form of molded shapes, it is used for pressure blast nozzles, wire-drawing dies, and bearing surfaces for gauges. For grinding and lapping applications, the available mesh sizes cover the range 240 to 800.	Boron carbide (Norbide®)

Table 3.2-20 Physical properties of carbides and carbide-based high-temperature refractories [2.4], cont.

IUPAC name (synonyms and common trade names)	Theoretical chemical formula, [CASRN], relative molecular mass ($^{12}C = 12.000$)	Crystal system, lattice parameters, *Strukturbericht* symbol, Pearson symbol, space group, structure type, Z	Density (ϱ, kg m^{-3})	Electrical resistivity (ρ, $\mu\Omega$ cm)	Melting point (°C)	Thermal conductivity (κ, W m^{-1} K^{-1})	Specific heat capacity (c_p, J kg^{-1} K^{-1})	Coefficient of linear thermal expansion (α, 10^{-6} K^{-1})
Chromium carbide	Cr$_7$C$_3$ 400.005	Hexagonal $a = 1389.02$ pm $c = 453.20$ pm	6992	109.0	1665	–	–	11.7
Chromium carbide	Cr$_3$C$_2$ [12012-35-0] 180.010	Orthorhombic $a = 282$ pm $b = 553$ pm $c = 1147$ pm D5$_{10}$, $oP20$, *Pbnm*, Cr$_3$C$_2$ type ($Z = 4$)	6680	75.0	1895	19.2	–	10.3
Diamond	C [7782-40-3] 12.011	Cubic $a = 356.683$ pm A4, $cF8$, $Fd3m$, diamond type ($Z = 8$)	3515.24	$>10^{16}$ (types I and IIa) $>10^3$ (type IIb)	3550	900 (type I) 2400 (type IIa)	–	2.16
Graphite	C [7782-42-5] 12.011	Hexagonal $a = 246$ pm $b = 428$ pm $c = 671$ pm A9, $hP4$, $P6_3/mmc$, graphite type ($Z = 4$)	2250	1385	3650	–	–	0.6–4.3
Hafnium monocarbide	HfC [12069-85-1] 190.501	Cubic $a = 446.0$ pm B1, $cF8$, $Fm3m$, rock salt type ($Z = 4$)	12670	45.0	3890–3950	22.15	–	6.3
Lanthanum dicarbide	LaC$_2$ [12071-15-7] 162.928	Tetragonal $a = 394.00$ pm $c = 657.20$ pm C11a, $tI6$, $I4/mmm$, CaC$_2$ type ($Z = 2$)	5290	68.0	2360–2438	–	–	12.1
Molybdenum hemicarbide	β-Mo$_2$C [12069-89-5] 203.891	Hexagonal $a = 300.20$ pm $c = 427.40$ pm L'3, $hP3$, $P6_3/mmc$, Fe$_2$N type ($Z = 1$)	9180	71.0	2687	–	29.4	7.8
Molybdenum monocarbide	MoC [12011-97-1] 107.951	Hexagonal $a = 290$ pm $c = 281$ pm B$_k$, $P6_3/mmc$, BN type ($Z = 4$)	9159	50.0	2577	–	–	5.76
Niobium hemicarbide	Nb$_2$C [12011-99-3] 197.824	Hexagonal $a = 312.70$ pm $c = 497.20$ pm L'3, $hP3$, $P6_3/mmc$, Fe$_2$N type ($Z = 1$)	7800	–	3090	–	–	–
Niobium monocarbide	NbC [12069-94-2] 104.917	Cubic $a = 447.71$ pm B1, $cF8$, $Fm3m$, rock salt type ($Z = 4$)	7820	51.1–74.0	3760	14.2	–	6.84

Table 3.2-20 Physical properties of carbides and carbide-based high-temperature refractories [2.4], cont.

Young's modulus (E, GPa)	Flexural strength (τ, MPa)	Compressive strength (α, MPa)	Vickers hardness HV (Mohs hardness HM)	Other physicochemical properties, corrosion resistance,[a] and uses	IUAPC name (synonyms and common trade names)
–	–	–	1336	Resists oxidation in the range 800–1000 °C. Corroded by the following molten metals: Ni and Zn.	Chromium carbide
386	–	1041	2650	Corroded by the following molten metals: Ni, Zn, Cu, Cd, Al, Mn, and Fe. Corrosion-resistant in molten Sn and Bi.	Chromium carbide
930	–	7000	8000 HK (HM 10)	Type I contains 0.1–0.2% N, type IIa is N-free, and type IIb is very pure, generally blue in color. Electrical insulator ($E_g = 7$ eV). Burns in oxygen.	Diamond
6.9	–	–	(HM 2)	High-temperature lubricant, used in crucibles for handling molten metals such as Mg, Al, Zn, Ga, Sb, and Bi.	Graphite
424	–	–	1870–2900	Dark, gray, brittle solid, the most refractory binary material known. Used in control rods in nuclear reactors and in crucibles for melting HfO_2 and other oxides. Corrosion-resistant to liquid metals such as Nb, Ta, Mo, and W. Severe oxidation in air above 1100–1400 °C, but stable up to 2000 °C in helium.	Hafnium monocarbide
–	–	–	–	Decomposed by H_2O.	Lanthanum dicarbide
221	–	n.a	1499 (HM > 7)	Gray powder. Wear-resistant film. Oxidized in air at 700–800 °C. Corroded in the following molten metals: Al, Mg, V, Cr, Mn, Fe, Ni, Cu, Zn, and Nb. Corrosion-resistant in molten Cd, Sn, and Ta.	Molybdenum hemicarbide
197	n.a.	n.a.	1800 (HM > 9)	Oxidized in air at 700–800 °C.	Molybdenum monocarbide
n.a.	n.a.	n.a.	2123		Niobium hemicarbide
340	n.a.	n.a.	2470 (HM > 9)	Lavender-gray powder, soluble in HF-HNO_3 mixture. Wear-resistant film, used for coating graphite in nuclear reactors. Oxidation in air becomes severe only above 1000 °C.	Niobium monocarbide

Table 3.2-20 Physical properties of carbides and carbide-based high-temperature refractories [2.4], cont.

IUPAC name (synonyms, and common trade names)	Theoretical chemical formula, [CASRN], relative molecular mass ($^{12}C = 12.000$)	Crystal system, lattice parameters, *Strukturbericht* symbol, Pearson symbol, space group, structure type (Z)	Density (ϱ, kg m^{-3})	Electrical resistivity (ρ, $\mu\Omega$ cm)	Melting point (°C)	Thermal conductivity (κ, W m^{-1} K^{-1})	Specific heat capacity (c_p, J kg^{-1} K^{-1})	Coefficient of linear thermal expansion (α, 10^{-6} K^{-1})
Silicon monocarbide (moissanite, Carbolon®, Crystolon®, Carborundum®)	α-SiC [409-21-2] 40.097	Hexagonal $a = 308.10$ pm $c = 503.94$ pm B4, hP4, P6$_3$/mmc, wurtzite type (Z = 2)	3160	4.1 × 10^5	2093 transformation temperature	42.5	690	4.3–4.6
Silicon monocarbide (Carbolon®, Crystolon®, Carborundum®)	β-SiC [409-21-2] 40.097	Cubic $a = 435.90$ pm B3, cF8, F43m, ZnS type (Z = 4)	3160	107–200	~2700, subl.	135	1205	4.5
Tantalum hemicarbide	Ta$_2$C [12070-07-4] 373.907	Hexagonal $a = 310.60$ pm $c = 493.00$ pm L'3, hP3, P6$_3$/mmc, Fe$_2$N type (Z = 2)	15 100	80.0	3327	–	–	–
Tantalum monocarbide	TaC [12070-06-3] 194.955	Cubic $a = 445.55$ pm B1, cF8, Fm3m, rock salt type (Z = 4)	14 800	30–42.1	3880	22.2	190	6.64–8.4
Thorium dicarbide	α-ThC$_2$ [12071-31-7] 256.060	Tetragonal $a = 585$ pm $c = 528$ pm C11a, tI6, I4mmm, CaC$_2$ type (Z = 2)	8960–9600	30.0	2655	23.9	–	8.46
Thorium monocarbide	ThC [12012-16-7] 244.089	Cubic $a = 534.60$ pm B1, cF8, Fm3m, rock salt type (Z=4)	10 670	25.0	2621	28.9	–	6.48
Titanium monocarbide	TiC [12070-08-5] 59.878	Cubic $a = 432.8$ pm B1, cF8, Fm3m, rock salt type (Z = 4)	4938	52.5	3140 ± 90	17–21	–	7.5–7.7
Tungsten hemicarbide	W$_2$C [12070-13-2] 379.691	Hexagonal $a = 299.82$ pm $c = 472.20$ pm L'3, hP3, P6$_3$/mmc, Fe$_2$N type (Z = 1)	17 340	81.0	2730	–	–	3.84
Tungsten monocarbide (Widia®)	WC [12070-12-1] 195.851	Hexagonal $a = 290.63$ pm $c = 283.86$ pm L'3, hP3, P6$_3$/mmc, Fe$_2$N type (Z = 1)	15 630	19.2	2870	121	–	6.9

Table 3.2-20 Physical properties of carbides and carbide-based high-temperature refractories [2.4], cont.

Young's modulus (E, GPa)	Flexural strength (τ, MPa)	Compressive strength (α, MPa)	Vickers hardness HV (Mohs hardness HM)	Other physicochemical properties, corrosion resistance,[a] and uses	IUPAC name (synonyms and common trade names)
386–414	–	500	2400–2500 (HM 9.2)	Semiconductor ($E_g = 3.03$ eV). Soluble in fused alkalimetal hydroxides.	Silicon monocarbide (moissanite, Carbolon®, Crystolon®, Carborundum®)
262–468	–	1000	2700–3350 (HM 9.5)	Green to bluish-black, iridescent crystals. Soluble in fused alkalimetal hydroxides. Abrasives made from this material are best suited to the grinding of low-tensile-strength materials such as cast iron, brass, bronze, marble, concrete, stone, glass, optical, structural, and wear-resistant components. Corroded by molten metals such as Na, Mg, Al, Zn, Fe, Sn, Rb, and Bi. Resistant to oxidation in air up to 1650 °C. Maximum operating temperature of 2000 °C in reducing or inert atmosphere.	Silicon monocarbide (Carbolon®, Crystolon®, Carborundum®)
–	–	–	1714–2000		Tantalum hemicarbide
364	–	–	1599–1800 (HM 9–10)	Golden-brown crystals, soluble in HF-HNO$_3$ mixture. Used in crucibles for melting ZrO$_2$ and similar oxides with high melting points. Corrosion-resistant to molten metals such as Ta, and Re. Readily corroded by liquid metals such as Nb, Mo, and Sn. Burning occurs in pure oxygen above 800 °C. Severe oxidation in air above 1100–1400 °C. Maximum operating temperature of 3760 °C under helium.	Tantalum monocarbide
–	–	–	600	$\alpha - \beta$ transition at 1427 °C and $\beta - \gamma$ transition at 1497 °C. Decomposed by H$_2$O with evolution of C$_2$H$_6$.	Thorium dicarbide
–	–	–	1000	Readily hydrolyzed in water, evolving C$_2$H$_6$.	Thorium monocarbide
310–462	–	1310	2620–3200 (HM 9–10)	Gray crystals. Superconducting at 1.1 K. Soluble in HNO$_3$ and aqua regia. Resistant to oxidation in air up to 450 °C. Maximum operating temperature 3000 °C under helium. Used in crucibles for handling molten metals such as Na, Bi, Zn, Pb, Sn, Rb, and Cd. Corroded by the following liquid metals: Mg, Al, Si, Ti, Zr, V, Nb, Ta, Cr, Mo, Mn, Fe, Co, and Ni. Attacked by molten NaOH.	Titanium monocarbide
421	–	–	3000	Black. Resistant to oxidation in air up to 700 °C. Corrosion-resistant to Mo.	Tungsten hemicarbide
710	–	530	2700 (HM > 9)	Gray powder, dissolved by HF-HNO$_3$ mixture. Cutting tools, wear-resistant semiconductor films. Corroded by the following molten metals: Mg, Al, V, Cr, Mn, Ni, Cu, Zn, Nb, and Mo. Corrosion-resistant to molten Sn.	Tungsten monocarbide (Widia®)

Table 3.2-20 Physical properties of carbides and carbide-based high-temperature refractories [2.4], cont.

IUPAC name (synonyms, and common trade names)	Theoretical chemical formula, [CASRN], relative molecular mass ($^{12}C = 12.000$)	Crystal system, lattice parameters, *Strukturbericht* symbol, Pearson symbol, space group, structure type (Z)	Density (ϱ, kg m^{-3})	Electrical resistivity (ρ, $\mu\Omega$ cm)	Melting point (°C)	Thermal conductivity (κ, W m^{-1} K^{-1})	Specific heat capacity (c_p, J kg^{-1} K^{-1})	Coefficient of linear thermal expansion (α, 10^{-6} K^{-1})
Uranium carbide	U$_2$C$_3$ [12076-62-9] 512.091	Cubic $a = 808.89$ pm D5c, cI40, I$\bar{4}$3d, Pu$_2$C$_3$ type (Z = 8)	12 880	–	1777	–	–	11.4
Uranium dicarbide	UC$_2$ [12071-33-9] 262.051	Tetragonal $a = 352.24$ pm $c = 599.62$ pm C11a, tI6, I4/mmm, CaC$_2$ type (Z = 2)	11 280	–	2350–2398	32.7	147	14.6
Uranium monocarbide	UC [12070-09-6] 250.040	Cubic $a = 496.05$ pm B1, cF8, Fm3m, rock salt type (Z = 4)	13 630	50.0	2370–2790	23.0	–	11.4
Vanadium hemicarbide	V$_2$C [2012-17-8] 113.89	Hexagonal $a = 286$ pm $c = 454$ pm L'3, hP3, P6$_3$mmc, Fe$_2$N type (Z = 2)	5750	–	2166	–	–	–
Vanadium monocarbide	VC [12070-10-9] 62.953	Cubic $a = 413.55$ pm B1, cF8, Fm3m, rock salt type (Z = 4)	5770	65.0–98.0	2810	24.8	–	4.9
Zirconium monocarbide	ZrC [12020-14-3] 103.235	Cubic $a = 469.83$ pm B1, cF8, Fm3m, rock salt type (Z = 4)	6730	68.0	3540–3560	20.61	205	6.82

[a] Corrosion data in molten salts from [2.9].

Table 3.2-20 Physical properties of carbides and carbide-based high-temperature refractories [2.4], cont.

Young's modulus (E, GPa)	Flexural strength (τ, MPa)	Compressive strength (α, MPa)	Vickers hardness HV (Mohs hardness HM)	Other physicochemical properties, corrosion resistance,[a] and uses	IUPAC name (synonyms and common trade names)
179–221	–	434	–		Uranium carbide
–	–	–	600	Transition from tetragonal to cubic at 1765 °C. Decomposed in H_2O, slightly soluble in alcohol. Used in microsphere pellets to fuel nuclear reactors.	Uranium dicarbide
172.4	–	351.6	750–935 (HM > 7)	Gray crystals with metallic appearance, reacts with oxygen. Corroded by the following molten metals: Be, Si, Ni, and Zr.	Uranium monocarbide
–	–	–	3000	Corroded by molten Nb, Mo, and Ta.	Vanadium hemicarbide
614	–	613	2090	Black crystals, soluble in HNO_3 with decomposition. Used in wear-resistant films and cutting tools. Resistant to oxidation in air up to 300 °C.	Vanadium monocarbide
345	–	1641	1830–2930 (HM > 8)	Dark gray, brittle solid, soluble in HF solutions containing nitrates or peroxide ions. Used in nuclear power reactors and in crucibles for handling molten metals such as Bi, Cd, Pb, Sn, and Rb, and molten zirconia (ZrO_2). Corroded by the following liquid metals: Mg, Al, Si, V, Nb, Ta, Cr, Mo, Mn, Fe, Co, Ni, and Zn. In air, oxidized rapidly above 500 °C. Maximum operating temperature of 2350 °C under helium.	Zirconium monocarbide

Table 3.2-21 Properties of carbides according to DIN EN 60672 [2.6]

			Designation				
			SSIC	SISIC	RSIC	NSIC	BC
Mechanical properties	Symbol	Units	Silicon carbide, sintered	Silicon carbide, silicon-infiltrated	Silicon carbide, recrystallized	Silicon carbide, nitride-bonded	Boron carbide
Open porosity		vol. %	0	0	0–15	–	0
Density, minimum	ϱ	Mg/m^3	3.08–3.15	3.08–3.12	2.6–2.8	2.82	2.50
Bending strength	σ_B	MPa	300–600	180–450	80–120	200	400
Young's modulus	E	GPa	370–450	270–350	230–280	150–240	390–440
Hardness	HV	100	25–26	14–25	25	–	30–40
Fracture toughness	K_{IC}	MPa $\sqrt{m}$	3.0–4.8	3.0–5.0	3.0	–	3.2–3.6
Electrical properties							
Resistivity at 20 °C	ρ_{20}	Ω m	10^3–10^4	2×10^1–10^3	–	–	–
Resistivity at 600 °C	ρ_{600}	Ω m	10	5	–	–	–
Thermal properties							
Average coefficient of thermal expansion at 30–600 °C	$\alpha_{30-1000}$	10^{-6} K^{-1}	4–4.8	4.3–4.8	4.8	4.5	6
Specific heat capacity at 30–100 °C	$c_{p,30-1000}$	J kg^{-1}K^{-1}	600–1000	650–1000	600–900	800–900	–
Thermal conductivity	λ_{30-100}	W m^{-1}K^{-1}	40–120	110–160	20	14–15	28
Thermal fatigue resistance		(Rated)	Very good	Very good	Very good	Very good	–
Typical maximum application temperature	T	°C	1400–1750	1380	1600	1450	700–1000

3.2.5.4 Nitrides

Nitrides are treated extensively in [2.1–3]. Some properties are listed in Tables 3.2-22 and 3.2-23.

Silicon Nitride

Si_3N_4 is the dominant nitride ceramic material because of its favorable combination of properties. The preparation of powders for the formation of dense silicon nitride materials requires the use of precursors. Four routes for the production of Si_3N_4 powders are used in practice: nitridation of silicon, chemical vapor deposition from $SiCl_4 + NH_3$, carbothermal reaction of SiO_2, and precipitation of silicon diimide $Si(NH)_2$ followed by decomposition. Table 3.2-24 gives examples of the properties of the resulting powders.

Table 3.2-22 Properties of nitrides according to DIN EN 60672 [2.6]

			Designation				
			SSN	RBSN	HPSN	SRBSN	AlN
Mechanical properties	Symbol	Units	Silicon nitride, sintered	Silicon nitride, reaction-bonded	Silicon nitride, hot-pressed	Silicon nitride, reaction-bonded	Aluminium nitride
Open porosity		vol. %	–	–	0	–	0
Density, minimum	ϱ	Mg/m^3	3–3.3	1.9–2.5	3.2–3.4	3.1–3.3	3.0
Bending strength	σ_B	MPa	700–1000	200–330	600–800	700–1200	200
Young's modulus	E	GPa	250–330	80–180	600–800	150–240	320
Hardness	HV	100	4–18	8–10	15–16	–	11
Fracture toughness	K_{IC}	MPa $\sqrt{m}$	5–8.5	1.8–4.0	6.0–8.5	3.0–6?	3.0
Electrical properties							
Resistivity at 20 °C	ρ_{20}	Ω m	10^{11}	10^{13}	10^{13}	–	10^{13}
Resistivity at 600 °C	ρ_{600}	Ω m	10^2	10^{10}	10^9	–	10^{12}
Thermal properties							
Average coefficient of thermal expansion at 30–600 °C	$\alpha_{30-1000}$	10^{-6} K^{-1}	2.5–3.5	2.1–3	3.1–3.3	3.0–3.4	4.5–5
Specific heat capacity at 30–100 °C	$c_{p,30-1000}$	J kg^{-1}K^{-1}	700–850	700–850	700–850	700–850	–
Thermal conductivity	λ_{30-100}	W m^{-1}K^{-1}	15–45	4–15	15–40	14–15	> 100
Thermal fatigue resistance		(Rated)	Very good	Very good	Very good	Very good	Very good
Typical maximum application temperature	T	°C	1250	1100	1400	1250	–

Table 3.2-23 Physical properties of nitrides and nitride-based high-temperature refractories [2.4]

IUPAC name (synonyms and common trade names)	Theoretical chemical formula, [CASRN], relative molecular mass ($^{12}C = 12.000$)	Crystal system, lattice parameters, *Strukturbericht* symbol, Pearson symbol, space group, structure type, Z	Density (ϱ, kg m^{-3})	Electrical resistivity (ρ, $\mu\Omega$ cm)	Melting point (°C)	Thermal conductivity (κ, W m^{-1} K^{-1})	Specific heat capacity (c_p, J kg^{-1} K^{-1})	Coefficient of linear thermal expansion (α, 10^{-6} K^{-1})
Aluminium mononitride	AlN [24304-00-5] 40.989	Hexagonal $a = 311.0$ pm $c = 497.5$ pm B4, hP4, P6$_3$mc, wurtzite type (Z = 2)	3050	10^{17}	2230	29.96	820	5.3
Beryllium nitride	α-Be$_3$N$_2$ [1304-54-7] 55.050	Cubic $a = 814$ pm D5$_3$, cI80, Ia3, Mn$_2$O$_3$ type (Z = 16)	2710	–	2200	–	1221	–
Boron mononitride	BN [10043-11-5] 24.818	Hexagonal $a = 250.4$ pm $c = 666.1$ pm B$_k$, hP8, P6$_3$/mmc, BN type (Z = 4)	2250	10^{19}	2730 (dec.)	15.41	711	7.54
Boron mononitride (Borazon®, CBN)	BN 24.818	Cubic $a = 361.5$ pm	3430	1900 (200°C)	1540	–	–	–
Chromium heminitride	Cr$_2$N [12053-27-9] 117.999	Hexagonal $a = 274$ pm $c = 445$ pm L'3, hP3, P6$_3$/mmc, Fe$_2$N type (Z = 1)	6800	76	1661	22.5	630	9.36
Chromium mononitride	CrN [24094-93-7] 66.003	Cubic $a = 415.0$ pm B1, cF8, Fm3m, rock salt type (Z = 4)	6140	640	1499 (dec.)	12.1	795	2.34
Hafnium mononitride	HfN [25817-87-2] 192.497	Cubic $a = 451.8$ pm B1, cF8, Fm3m, rock salt type (Z = 4)	13 840	33	3310	21.6	210	6.5
Molybdenum heminitride	Mo$_2$N [12033-31-7] 205.887	Cubic $a = 416$ pm L'1, cP5, Pm3m, Fe$_4$N type (Z = 2)	9460	19.8	760–899	17.9	293	6.12
Molybdenum mononitride	MoN [12033-19-1] 109.947	Hexagonal $a = 572.5$ pm $c = 560.8$ pm B$_h$, hP2, P6/mmm, WC type (Z = 1)	9180	–	1749	–	–	–
Niobium mononitride	NbN [24621-21-4] 106.913	Cubic $a = 438.8$ pm B1, cF8, Fm3m, rock salt type (Z = 4)	8470	78	2575	3.63	–	10.1
Silicon nitride	β-Si$_3$N$_4$ [12033-89-5] 140.284	Hexagonal $a = 760.8$ pm $c = 291.1$ pm P6/3m	3170	10^6	1850	28	713	2.25

Table 3.2-23 Physical properties of nitrides and nitride-based high-temperature refractories [2.4], cont.

Young's modulus (E, GPa)	Flexural strength (τ, MPa)	Compressive strength (α, MPa)	Vickers hardness HV (Mohs hardness HM)	Other physicochemical properties, corrosion resistance,[a] and uses	IUPAC name (synonyms and common trade names)
346	–	2068	1200 (HM 9–10)	Insulator ($E_g = 4.26$ eV). Decomposed by water, acids, and alkalis to Al(OH)$_3$ and NH$_3$. Used in crucibles for GaAs crystal growth.	Aluminium mononitride
–	–	–	–	Hard white or grayish crystals. Oxidized in air above 600 °C. Slowly decomposed in water, quickly in acids and alkalis, with evolution of NH$_3$.	Beryllium nitride
85.5	–	310	230 (HM 2.0)	Insulator ($E_g = 7.5$ eV). Used in crucibles for molten metals such as Na, B, Fe, Ni, Al, Si, Cu, Mg, Zn, In, Bi, Rb, Cd, Ge, and Sn. Corroded by these molten metals: U, Pt, V, Ce, Be, Mo, Mn, Cr, V, and Al. Attacked by the following molten salts: PbO$_2$, Sb$_2$O$_3$, Bi$_2$O$_3$, KOH, and K$_2$CO$_3$. Used in furnace insulation, diffusion masks, and passivation layers.	Boron mononitride
–	–	7000	4700–5000 (HM 10)	Tiny reddish to black grains. Used as an abrasive for grinding tool and die steels and high-alloy steels when chemical reactivity of diamond is a problem.	Boron mononitride (Borazon®, CBN)
–	–	–	1200–1571		Chromium heminitride
–	–	–	1090		Chromium mononitride
–	–	–	1640 (HM > 8–9)	Most refractory of all nitrides.	Hafnium mononitride
–	–	–	1700	Phase transition at 5.0 K.	Molybdenum heminitride
–	–	–	650		Molybdenum mononitride
–	–	–	1400 (HM > 8)	Dark gray crystals. Transition temperature 15.2 K. Insoluble in HCl, HNO$_3$, and H$_2$SO$_4$, but attacked by hot caustic solutions, lime, or strong alkalis, evolving NH$_3$.	Niobium mononitride
55	–	–	(HM > 9)		Silicon nitride

Table 3.2-23 Physical properties of nitrides and nitride-based high-temperature refractories [2.4], cont.

IUPAC name (synonyms and common trade names)	Theoretical chemical formula, [CASRN], relative molecular mass ($^{12}C = 12.000$)	Crystal system, lattice parameters, Strukturbericht symbol, Pearson symbol, space group, structure type, Z	Density (ϱ, kg m^{-3})	Electrical resistivity (ρ, $\mu\Omega$ cm)	Melting point (°C)	Thermal conductivity (κ, W m^{-1} K^{-1})	Specific heat capacity (c_p, J kg^{-1} K^{-1})	Coefficient of linear thermal expansion (α, 10^{-6} K^{-1})
Silicon nitride (Nitrasil®)	α-Si$_3$N$_4$ [12033-89-5] 140.284	Hexagonal $a = 775.88$ pm $c = 561.30$ pm $P3\mathit{1}c$	3184	10^{19}	1900 (sub.)	17	700	2.5–3.3
Tantalum heminitride	Ta$_2$N 375.901	Hexagonal $a = 306$ pm $c = 496$ pm $L'3, hP3, P6_3/mmc$, Fe$_2$N type ($Z = 1$)	15 600	263	2980	10.04	126	5.2
Tantalum mononitride (ϵ)	TaN [12033-62-4] 194.955	Hexagonal $a = 519.1$ pm $c = 290.6$ pm	13 800	128–135	3093	8.31	210	3.2
Thorium mononitride	ThN [12033-65-7] 246.045	Cubic $a = 515.9$ B1, $cF8, Fm3m$, rock salt type ($Z = 4$)	11 560	20	2820	–	–	7.38
Thorium nitride	Th$_2$N$_3$ [12033-90-8]	Hexagonal $a = 388$ pm $c = 618$ pm $D5_2, hP5, \bar{3}ml$, La$_2$O$_3$ type ($Z = 1$)	10 400	–	1750	–	–	–
Titanium mononitride	TiN [25583-20-4] 61.874	Cubic $a = 424.6$ pm B1, $cF8, Fm3m$, rock salt type ($Z = 4$)	5430	21.7	2930 (dec.)	29.1	586	9.35
Tungsten dinitride	WN$_2$ [60922-26-1] 211.853	Hexagonal $a = 289.3$ pm $c = 282.6$ pm	7700	–	600 (dec.)	–	–	–
Tungsten heminitride	W$_2$N [12033-72-6] 381.687	Cubic $a = 412$ pm $L'1, cP5, Pm3m$, Fe$_4$N type ($Z = 2$)	17 700	–	982	–	–	–
Tungsten mononitride	WN [12058-38-7]	Hexagonal	15 940	–	593	–	–	–
Uranium nitride	U$_2$N$_3$ [12033-83-9] 518.259	Cubic $a = 1070$ pm $D5_3, c180, Ia3$, Mn$_2$O$_3$ type ($Z = 16$)	11 240	–	–	–	–	–
Uranium mononitride	UN [25658-43-9] 252.096	Cubic $a = 489.0$ pm B1, $cF8, Fm3m$, rock salt type ($Z = 4$)	14 320	208	2900	12.5	188	9.72
Vanadium mononitride	VN [24646-85-3] 64.949	Cubic $a = 414.0$ pm B1, $cF8, Fm3m$, rock salt type ($Z = 4$)	6102	86	2360	11.25	586	8.1
Zirconium mononitride	ZrN [25658-42-8] 105.231	Cubic $a = 457.7$ pm B1, $cF8, Fm3m$, rock salt type ($Z = 4$)	7349	13.6	2980	20.90	377	7.24

Table 3.2-23 Physical properties of nitrides and nitride-based high-temperature refractories [2.4], cont.

Young's modulus (E, GPa)	Flexural strength (τ, MPa)	Compressive strength (α, MPa)	Vickers hardness HV (Mohs hardness HM)	Other physicochemical properties, corrosion resistance,[a] and uses	IUAPC name (synonyms and common trade names)
304	–	–	(HM > 9)	Gray amorphous powder or crystals. Corrosion-resistant to molten metals such as Al, Pb, Zn, Cd, Bi, Rb, and Sn, and molten salts NaCl-KCl, NaF, and silicate glasses. Corroded by molten Mg, Ti, V, Cr, Fe, Co, cryolite, KOH, and Na_2O.	Silicon nitride (Nitrasil®)
–	–	–	3200	Decomposed by KOH with evolution of NH_3.	Tantalum heminitride
–	–	–	1110 (HM > 8)	Bronze-colored or black crystals. Transition temperature 1.8 K. Insoluble in water, slowly attacked by aqua regia, HF, and HNO_3.	Tantalum mononitride (ϵ)
–	–	–	600	Gray solid. Slowly hydrolyzed by water.	Thorium mononitride
–	–	–	–		Thorium nitride
248	–	972	1900 (HM 8–9)	Bronze-colored powder. Transition temperature 4.2 K. Corrosion-resistant to molten metals such as Al, Pb, Mg, Zn, Cd, and Bi. Corroded by molten Na, Rb, Ti, V, Cr, Mn, Sn, Ni, Cu, Fe, and Co. Dissolved by boiling aqua regia; decomposed by boiling alkalis, evolving NH_3.	Titanium mononitride
–	–	–	–	Brown crystals.	Tungsten dinitride
–	–	–	–	Gray crystals.	Tungsten heminitride
–	–	–	–	Gray solid. Slowly hydrolyzed by water.	Tungsten mononitride
–	–	–	–		Uranium nitride
149	–	–	455		Uranium mononitride
–	–	–	1520 (HM 9–10)	Black powder. Transition temperature 7.5 K. Soluble in aqua regia.	Vanadium mononitride
–	–	979	1480 (HM > 8)	Yellow solid. Transition temperature 9 K. Corrosion-resistant to steel, basic slag, and cryolite, and molten metals such as Al, Pb, Mg, Zn, Cd, and Bi. Corroded by molten Be, Na, Rb, Ti, V, Cr, Mn, Sn, Ni, Cu, Fe, and Co. Soluble in concentrated HF, slowly soluble in hot H_2SO_4.	Zirconium mononitride

[a] Corrosion data in molten salts from [2.9].

Table 3.2-24 Characteristics of Si_3N_4 powders processed by different preparation techniques [2.3]

	Technique						
	Nitridation of Si		Chemical vapor deposition		Carbothermal reduction	Diimide precipitation	
Sample no.	1	2	1	2		1	2
Specific surface area (m²/g)	23	11	4	10	10	11	13
O (wt%)	1.4	1.0	1.0	3.0	2.0	1.4	1.5
C (wt%)	0.2	0.25	–	–	0.9	0.1	0.1
Fe, Al, Ca (wt%)	0.07	0.4	0.005	0.005	0.22	0.01	0.015
Other impurities (wt%)			Cl 0.04 Mo+Ti 0.02			Cl 0.1	Cl 0.005
Crystallinity (%)	100	100	60	0	100	98	–
$\alpha/(\alpha+\beta)$ (%)	95	92	95	–	98	86	95
Morphology[a]	E	E	E+R	E+R	E+R	E	E

[a] E, equiaxed; R, rod-like.

Table 3.2-25 Physical properties of silicides and silicide-based high-temperature refractories [2.4]

IUPAC name	Theoretical chemical formula, [CASRN], relative molecular mass ($^{12}C = 12.000$)	Crystal system, lattice parameters, *Strukturbericht* symbol, Pearson symbol, space group, structure type, Z	Density (ϱ, kg m^{-3})	Electrical resistivity (ρ, $\mu\Omega$ cm)	Melting point (°C)	Thermal conductivity (κ, W m^{-1} K^{-1})	Specific heat capacity (c_p, J kg^{-1} K^{-1})	Coefficient of linear thermal expansion (α, 10^{-6} K^{-1})
Chromium disilicide	$CrSi_2$ [12018-09-6] 108.167	Hexagonal a = 442 pm c = 635 pm C40, hP9, P6$_2$22, CrSi$_2$ type (Z = 3)	4910	1400	1490	106	–	13.0
Chromium silicide	Cr_3Si [12018-36-9] 184.074	Cubic a = 456 pm A15, cP8, Pm3n, Cr$_3$Si type (Z = 2)	6430	45.5	1770	–	–	10.5
Hafnium disilicide	$HfSi_2$ [12401-56-8] 234.66	Orthorhombic a = 369 pm b = 1446 pm c = 346 pm C49, oC12, Cmcm, ZrSi$_2$ type (Z = 4)	8030	–	1699	–	–	–
Molybdenum disilicide	$MoSi_2$ [12136-78-6] 152.11	Tetragonal a = 319 pm c = 783 pm C11b, tI6, I4/mmm, MoSi$_2$ type (Z = 2)	6260	21.5	1870	58.9	–	8.12
Niobium disilicide	$NbSi_2$ [12034-80-9] 149.77	Hexagonal a = 479 pm c = 658 pm C40, hP9, P6$_2$22, CrSi$_2$ type (Z = 3)	5290	50.4	2160	–	–	–
Tantalum disilicide	$TaSi_2$ [12039-79-1] 237.119	Hexagonal a = 477 pm c = 655 pm C40, hP9, P6$_2$22, CrSi$_2$ type (Z = 3)	9140	8.5	2299	–	–	8.8–9.54

3.2.5.5 Silicides

Silicides, being compounds of silicon with metals, mostly show a metallic luster. Like intermetallic phases the silicides of a metal may occur in different stoichiometric variants, e.g. Ca_2Si, Ca_5Si_3, and $CaSi$. Silicides of the non-noble metals are unstable in contact with water and oxidizing media. Silicides of transition metals are highly oxidation-resistant. Silicides are used as ceramic materials mainly in high-temperature applications. $MoSi_2$ is used in resistive heating elements. Some properties are listed in Table 3.2-25.

Table 3.2-25 Physical properties of silicides and silicide-based high-temperature refractories [2.4], cont.

Young's modulus (E, GPa)	Flexural strength (τ, MPa)	Compressive strength (α, MPa)	Vickers hardness HV (Mohs hardness HM)	Other physicochemical properties, corrosion resistance and uses	IUPAC name
–	–	–	1000–1130		Chromium disilicide
–	–	–	1005		Chromium silicide
–	–	–	865–930		Hafnium disilicide
407	–	2068–2415	1260	The compound is thermally stable in air up to 1000 °C. Corrosion-resistant to molten metals such as Zn, Pd, Ag, Bi, and Rb. It is corroded by the following liquid metals: Mg, Al, Si, V, Cr, Mn, Fe, Ni, Cu, Mo, and Ce.	Molybdenum disilicide
–	–	–	1050		Niobium disilicide
–	–	–	1200–1600	Corroded by molten Ni.	Tantalum disilicide

Table 3.2-25 Physical properties of silicides and silicide-based high-temperature refractories [2.4], cont.

IUPAC name	Theoretical chemical formula, [CASRN], relative molecular mass ($^{12}C = 12.000$)	Crystal system, lattice parameters, Strukturbericht symbol, Pearson symbol, space group, structure type, Z	Density (ϱ, kg m^{-3})	Electrical resistivity (ρ, $\mu\Omega$ cm)	Melting point (°C)	Thermal conductivity (κ, W m^{-1} K^{-1})	Specific heat capacity (c_p, J kg^{-1} K^{-1})	Coefficient of linear thermal expansion (α, 10^{-6} K^{-1})
Tantalum silicide	Ta$_5$Si$_3$ [12067-56-0] 988.992	Hexagonal	13060	–	2499	–	–	–
Thorium disilicide	ThSi$_2$ [12067-54-8] 288.209	Tetragonal $a = 413$ pm $c = 1435$ pm Cc, tI12, I4amd, ThSi$_2$ type (Z = 4)	7790	–	1850	–	–	–
Titanium disilicide	TiSi$_2$ [12039-83-7] 104.051	Orthorhombic $a = 360$ pm $b = 1376$ pm $c = 360$ pm C49, oC12, Cmcm, ZrSi$_2$ type (Z = 4)	4150	123	1499	–	–	10.4
Titanium trisilicide	Ti$_5$Si$_3$ [12067-57-1] 323.657	Hexagonal $a = 747$ pm $c = 516$ pm D8$_8$, hP16, P6$_3$mcm, Mn$_5$Si$_3$ type (Z = 2)	4320	55	2120	–	–	110
Tungsten disilicide	WSi$_2$ [12039-88-2] 240.01	Tetragonal $a = 320$ pm $c = 781$ pm C11b, tI6, I4/mmm, MoSi$_2$ type (Z = 2)	9870	33.4	2165	–	–	8.28
Tungsten silicide	W$_5$Si$_3$ [12039-95-1] 1003.46		12210	–	2320	–	–	–
Uranium disilicide	USi$_2$ 294.200	Tetragonal $a = 397$ pm $c = 1371$ pm Cc, tI12, I4/amd, ThSi$_2$ type (Z = 4)	9250	–	1700	–	–	–
Uranium silicide	β-U$_3$Si$_2$ 770.258	Tetragonal $a = 733$ pm $c = 390$ pm D5a, tP10, P4/mbm, U$_3$Si$_2$ type (Z = 2)	12200	150	1666	14.7	–	14.8
Vanadium disilicide	VSi$_2$ [12039-87-1] 107.112	Hexagonal $a = 456$ pm $c = 636$ pm C40, hP9, P6$_2$22, CrSi$_2$ type (Z = 3)	5100	9.5	1699	–	–	11.2
Vanadium silicide	V$_3$Si [12039-76-8] 147.9085	Cubic $a = 471$ pm A15, cP8, Pm3n, Cr$_3$Si type (Z = 2)	5740	203	1732	–	–	8.0
Zirconium disilicide	ZrSi$_2$ [12039-90-6] 147.395	Orthorhombic $a = 372$ pm $b = 1469$ pm $c = 366$ pm C49, oC12, Cmcm, ZrSi$_2$ type (Z = 4)	4880	161	1604	–	–	8.6

Table 3.2-25 Physical properties of silicides and silicide-based high-temperature refractories [2.4], cont.

Young's modulus (E, GPa)	Flexural strength (τ, MPa)	Compressive strength (α, MPa)	Vickers hardness HV (Mohs hardness HM)	Other physicochemical properties, corrosion resistance, and uses	IUPAC name
–	–	–	1200–1500	The compound is thermally stable in air up to 400 °C.	Tantalum silicide
–	–	–	1120	Corrosion-resistant to molten Cu, while corroded by molten Ni.	Thorium disilicide
–	–	–	890–1039		Titanium disilicide
–	–	–	986		Titanium trisilicide
–	–	–	1090		Tungsten disilicide
–	–	–	770	Corroded by molten Ni.	Tungsten silicide
–	–	–	700		Uranium disilicide
77.9	–	–	796		Uranium silicide
–	–	–	1400		Vanadium disilicide
–	–	–	1500		Vanadium silicide
–	–	–	1030–1060		Zirconium disilicide

References

2.1 L. E. Toth: *Transition Metals, Carbides and Nitrides* (Academic Press, New York 1971)

2.2 R. Freer: *The Physics and Chemistry of Carbides, Nitrides and Borides* (Kluwer, Boston 1989)

2.3 M. V. Swain (Ed.): Structure and properties of ceramics. In: *Materials Science and Technology*, Vol. 11 (Verlag Chemie, Weinheim 1994)

2.4 F. Cardarelli: *Materials Handbook* (Springer, London 2000)

2.5 J. R. Davis (Ed.): *Heat-Resistant Materials*, ASM Specialty Handbook (ASM International, Materials Park 1997)

2.6 Verband der keramischen Industrie: *Brevier Technische Keramik* (Fahner Verlag, Lauf 1999) (in German)

2.7 P. Otschick (Ed.): *Langzeitverhalten von Funktionskeramiken* (Werkstoff-Informationsgesellschaft, Frankfurt 1997) (in German)

2.8 G. V. Samsonov: *The Oxides Handbook* (Plenum, New York 1974)

2.9 G. Geirnaert: Céamiques et mbaux liquides: Compatibilité et angles de mouillages, Bull. Soc. Fr. Ceram **106**, 7 (1970)

2.10 V. I. Matkovich (Ed.): *Boron and Refractory Borides* (Springer, Berlin, Heidelberg 1977)

3.3. Polymers

The physical properties of polymers depend not only on the kind of material but also on the molar mass, the molar-mass distribution, the kind of branching, the degree of branching, the crystallinity (amorphous or crystalline), the tacticity, the end groups, any superstructure, and any other kind of molecular architecture. In the case of copolymers, the physical properties are additionally influenced by the type of arrangement of the monomers (statistical, random, alternating, periodic, block, or graft). Furthermore, the properties of polymers are influenced if they are mixed with other polymers (polymer blends), with fibers (glass fibers, carbon fibers, or metal fibers), or with other fillers (cellulose, inorganic materials, or organic materials).

The tables and figures include the physical and physicochemical properties of those polymers, copolymers, and polymer blends which are widely used for scientific applications and in industry. The figures include mainly the following physical properties: stress versus strain, viscosity versus shear rate, and creep modulus versus time. However, other physical properties are also included. Additionally, the most relevant applications of the materials are given.

3.3.1	**Structural Units of Polymers**	480
3.3.2	**Abbreviations**	482
3.3.3	**Tables and Figures**	483
	3.3.3.1 Polyolefines	483
	3.3.3.2 Vinyl Polymers	489
	3.3.3.3 Fluoropolymers	492
	3.3.3.4 Polyacrylics and Polyacetals	497
	3.3.3.5 Polyamides	501
	3.3.3.6 Polyesters	503
	3.3.3.7 Polysulfones and Polysulfides	506
	3.3.3.8 Polyimides and Polyether Ketones	508
	3.3.3.9 Cellulose	509
	3.3.3.10 Polyurethanes	511
	3.3.3.11 Thermosets	512
	3.3.3.12 Polymer Blends	515
References		522

The tables and figures include the physical and physicochemical properties of the most important polymers, copolymers, and polymer blends. "Most important" here means that these materials are widely used for scientific applications and in industry. The values in the main tables are given for room temperature, that is, $\approx 25\,°\mathrm{C}$; otherwise, the temperature is given in parentheses. The tables and figures include the following physical properties:

Melting temperature T_m: heating rate 10 K/min (ISO 11357).

Enthalpy of fusion ΔH_u: the amount of enthalpy (given per monomer unit of the polymer) needed for the transition of the polymer from the solid state to the molten state.

Entropy of fusion ΔS_u: amount of entropy (given per monomer unit of the polymer) which is needed for the transition of a polymer from the solid state to the molten state.

Heat capacity $c_\mathrm{p} = (\partial H/\partial T)_p \approx \Delta H/\Delta T$; $\Delta H =$ quantity of heat per mass unit, $\Delta T =$ temperature increase.

Enthalpy of combustion ΔH_c: amount of enthalpy released in flaming combustion per unit mass of the polymer.

Glass transition temperature T_g: heating rate 10 K/min (ISO 11357).

Vicat softening temperature : $T_\mathrm{V} 10/50$, force 10 N, heating rate 50 K/h; $T_\mathrm{V} 50/50$, force 50 N, heating rate 50 K/h (ISO 306).

Thermal conductivity λ: $dq/dt = A\lambda\,dT/dx$; dq/dt = heat flux, A = area, dT/dx = temperature gradient.
Density $\varrho = m/V$ (ISO 1183).
Coefficient of expansion $\alpha = (1/V_0)(\partial V/\partial T)_p$: $T = 23\text{--}55\,°\text{C}$ (ISO 11359).
Compressibility $\kappa = -(1/V)(\partial V/\partial p)_T$.
Elastic modulus $E = \sigma/\varepsilon$ (σ = stress, ε = strain (elongation)); elongation rate 1 mm/min (ISO 527).
Shear modulus $G = \tau/\gamma$ (τ = shear stress, γ = shear angle).
Poisson's ratio $\mu = 0.5[1 - (E/\sigma)(\Delta V/V)]$; $\Delta V/V$ = relative volume change.
Stress at yield σ_y, strain (elongation) at yield ε_y: see Fig. 3.3-1; elongation rate 50 mm/min (ISO 527).
Stress at 50% strain (elongation) σ_{50}: see Fig. 3.3-1; elongation rate 50 mm/min (ISO 527).
Stress at fracture σ_b, strain (elongation) at fracture ε_b: see Fig. 3.3-1; elongation rate 5 mm/min (ISO 527).
Impact strength, and notched impact strength (Charpy) (ISO 179).
Sound velocity v_s, longitudinal (long) and transverse (trans).
Shore hardness A, D (ISO 868).
Volume resistivity ρ_e, surface resistivity σ_e: contact electrodes, voltage 500 V (DIN 0303 T30, ISO 93, IEC 60093).
Electric strength E_B: specimen of thickness 1.0 ± 0.1 mm (ISO 10350, IEC 60243).
Relative permittivity ε_r, dielectric loss (dissipation factor) $\tan\delta$ (IEC 60250).
Refractive index n_D, temperature coefficient of refractive index dn_D/dT.
Steam permeation: 20–25 °C, 85% relative humidity gradient (DIN 53122, ISO 15106).
Gas permeation: 20–25 °C, reduced to 23 °C, 1 bar (ISO 2556, DIN 53380, ISO 15105).
Melt-viscosity–molar-mass relation.
Viscosity–molar-mass relation: $[\eta] = KM^a$ means $[\eta]/[\eta_0] = K(M/M_0)^a$, where $[\eta_0] = 1\,\text{cm}^3/\text{g}$, $M_0 = 1\,\text{g/mol}$, and $[\eta]$ = intrinsic viscosity number at concentration $C = 0\,\text{g/cm}^3$ [3.1] (DIN 53726, DIN 53727).
Stress $\sigma(\varepsilon, T)$; ε = strain (elongation), T = temperature (ISO 527).
Viscosity $\eta(d\gamma/dt, T)$; $d\gamma/dt$ = shear rate, T = temperature (ISO 11443).
Creep modulus $E_{tc}(t, p, T)$; t = time, p = pressure, T = temperature; $E_{tc} = \sigma_{tc}/\varepsilon(t)$ (σ_{tc} = creep stress, $\varepsilon(t)$ = creep strain (creep elongation)); strain $\leq 0.5\%$ (ISO 899).

For selected polymers, the temperature dependence of some physical properties is given. Additionally, the most relevant applications of the materials are given. The tables and figures include the physical properties given in the table below (see [3.1–3]).

As the physical and physicochemical properties of each polymer vary with its molecular architecture, the tables show the ranges of the physical and physicochemical properties, whereas the diagrams show the functional relationships for a typical species of the polymer, copolymer, or polymer blend. The table on page 479 shows the selected 77 polymers, copolymers and polymer blends.

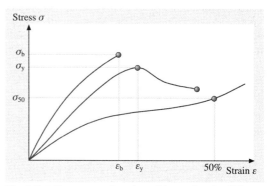

Fig. 3.3-1 Stress σ as a function of the strain ε for different kinds of polymers (see page 478)

3.3.3.1 Polyolefines

Polyolefines I

Polyethylene: high density HDPE, medium density MDPE, low density LDPE, linear low density LLDPE, ultra high molecular weight UHMWPE

Polyolefines II

Poly(ethylene-co-vinylacetate) EVA, Polyethylene ionomer EIM, Cycloolefine copolymer COC [Poly(ethylene-co-norbornene)], Poly(ethylene-co-acrylic acid) EAA

Polyolefines III

Polypropylene PP, Polybutene-1 PB, Polyisobutylene PIB, Poly(4-methylpentene-1) PMP

3.3.3.2 Vinylpolymers

Vinylpolymers I

Polystyrene PS, Poly(styrene-co-butadiene) SB, Poly(styrene-co-acrylonitrile) SAN

Vinylpolymers II

Poly(vinyl carbazole) PVK, Poly(acrylonitrile-co-butadiene-co-styrene) ABS, Poly(acrylonitrile-co-styrene-co-acrylester) ASA

Vinylpolymers III

Poly(vinyl chloride): unplastisized PVC-U, plastisized (75/25) PVC-P1, plastisized (60/40) PVC-P2

3.3.3.3 Fluoropolymers

Polytetrafluoroethylene PTFE, Polychlorotrifluoroethylene PCTFE, Poly(tetrafluoroethylene-co-hexafluoropro- pylene) FEP, Poly(ethylene-co-tetrafluoroethylene) ETFE, Poly(ethylene-co-chlorotrifluoroethylene), ECTFE

3.3.3.4 Polyacrylics, Polyacetals

Poly(methyl methacrylate) PMMA; Poly(oxymethylene) POM-H, Poly(oxymethylene-co-ethylene) POM-R

3.3.3.5 Polyamides

Polyamide 6 PA6, Polyamide 66 PA66, Polyamide 11 PA11, Polyamide 12 PA12, Polyamide 610 PA610

3.3.3.6 Polyesters

Polycarbonate PC, Poly(ethylene terephthalate) PET, Poly(butylene terephthalate) PBT, Poly(phenylene ether) PPE

3.3.3.7 Polysulfones, Polysulfides

Polysulfon PSU, Poly(phenylene sulfide) PPS, Poly(ether sulfone) PES

3.3.3.8 Polyimides, Polyether ketones

Poly(amide imide), PAI; Poly(ether imide), PEI; Polyimide, PI; Poly(ether ether ketone), PEEK

3.3.3.9 Cellulose

Cellulose acetate CA, Cellulose propionate CP, Cellulose acetobutyrate CAB, Ethyl cellulose EC, Vulcanized fiber VF

3.3.3.10 Polyurethanes

Polyurethane PUR, Thermoplastic polyurethane elastomer TPU

3.3.3.11 Thermosets

Thermosets I

Phenol formaldehyde PF, Urea formaldehyde UF, Melamine formaldehyde MF

Thermosets II

Unsaturated polyester UP, Diallylphthalat DAP, Silicone resin SI, Epoxy resin EP

3.3.3.12 Polymer Blends

Polymer Blends I

Polypropylene + Ethylene/propylene/diene-rubber PP + EPDM, Poly(acrylonitrile-co-butadiene-co-styrene) + Polycarbonate ABS + PC, Poly(acrylonitrile-co-butadiene-co-styrene) + Polyamide ABS + PA, Poly(acrylonitrile-co-butadiene-co-acrylester) + Polycarbonate ASA + PC

Polymer Blends II
Poly(vinyl chloride) + Poly(vinylchloride-co-acrylate) PVC + VC/A, Poly(vinyl chloride) + chlorinated Polyethylene PVC + PE-C, Poly(vinyl chloride) + Poly(acrylonitrile-co-butadiene-co-acrylester) PVC + ASA

Polymer Blends III
Polycarbonate + Poly(ethylene terephthalate) PC + PET, Polycarbonate + Liquid crystal polymer PC + LCP, Polycarbonate + Poly(butylene terephthalate) PC + PBT, Poly(ethylene terephthalate) + Polystyrene PET + PS, Poly(butylene terephthalate) + Polystyrene PBT + PS

Polymer Blends IV
Poly(butylene terephthalate) + Poly(acrylonitrile-co-butadiene-co-acrylester) PBT + ASA, Polysulfon + Poly(acrylonitrile-co-butadiene-co-styrene) PSU + ABS, Poly(phenylene ether) + Poly(styrene-co-butadiene) PPE + SB, Poly(phenylene ether) + Polyamide 66 PPE + PA66, Poly(phenylene ether) + Polystyrene PPE + PS

3.3.1 Structural Units of Polymers

The polymers given in this chapter are divided into polyolefines, vinyl polymers, fluoropolymers, polyacrylics, polyacetals, polyamides, polyesters, polysulfones, polysulfides, polyimides, polyether ketones, cellulose, polyurethanes, and thermosets. The structural units of the polymers are as follows:

Polyolefines

Polyethylene, PE: $-CH_2-CH_2-$

Polypropylene, PP: $-CH_2-CH(CH_3)-$

Poly(butene-1), PB: $-CH_2-CH(CH_2CH_3)-$

Poly(isobutylene), PIB: $-CH_2-C(CH_3)_2-$

Poly(4-methylpentene-1), PMP: $-CH_2-CH(CH_2-CH(CH_3)_2)-$

Polynorbornene: cyclopentane–CH=CH–

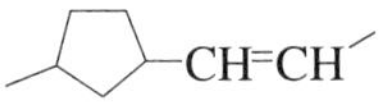

Poly(1,4-butadiene), BR: $-CH_2-CH=CH-CH_2-$

Vinyl Polymers

Polystyrene, PS: $-CH_2-CH(C_6H_5)-$

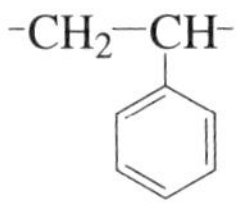

Poly(acrylonitrile), PAN: $-CH_2-CH(CN)-$

Poly(vinyl acetate), PVAC: $-CH_2-CH(O-CO-CH_3)-$

Poly(vinyl chloride), PVC: $-CH_2-CH(Cl)-$

Poly(vinyl carbazole), PVK: $-CH_2-CH(N(C_6H_5)_2)-$

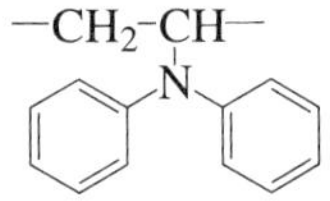

Fluoropolymers

Poly(tetrafluoroethylene), PTFE: $-CF_2-CF_2-$

Poly(chlorotrifluoroethylene), PCTFE: $-CFCl-CF_2-$

Poly(hexafluoropropylene): $-CF_2-CF(CF_3)-$

Polyacrylics and Polyacetals

Poly(methyl methacrylate), PMMA: $-CH_2-C(CH_3)(COOCH_3)-$

Poly(acrylic acid), PAA: $-CH_2-CH(COO^-)-$

Poly(oxymethylene), POM: $-CH_2-O-$

Polyamides

Polyamide 6, PA6: $-CO-(CH_2)_5-NH-$

Polyamide 66, PA66:
$-NH-(CH_2)_6-NH-CO-(CH_2)_4-CO-$

Polyamide 11, PA11: $-CO-(CH_2)_{10}-NH-$

Polyamide 12, PA12: $-CO-(CH_2)_{11}-NH-$

Polyamide 610, PA610:
$-NH-(CH_2)_6-NH-CO-(CH_2)_8-CO-$

Polyamide 612, PA612:
$-NH-(CH_2)_6-NH-CO-(CH_2)_{10}-CO-$

Polyesters
Polycarbonate, PC:

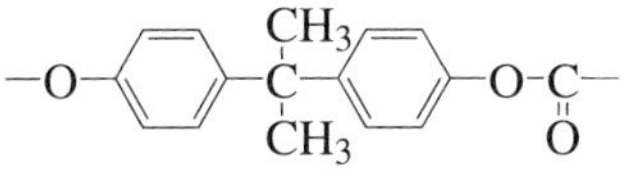

Poly(ethylene terephthalate), PET:

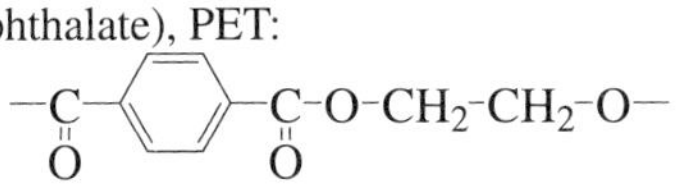

Poly(butylene terephthalate), PBT:

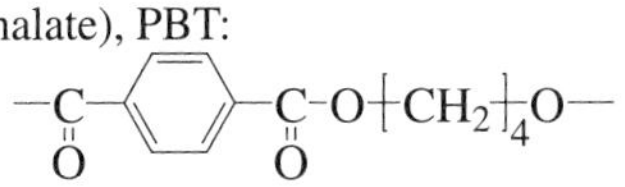

Poly(phenylene ether), PPE:

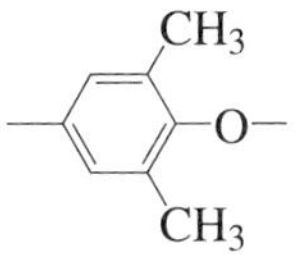

Polysulfones and Polysulfides
Polysulfone, PSU:

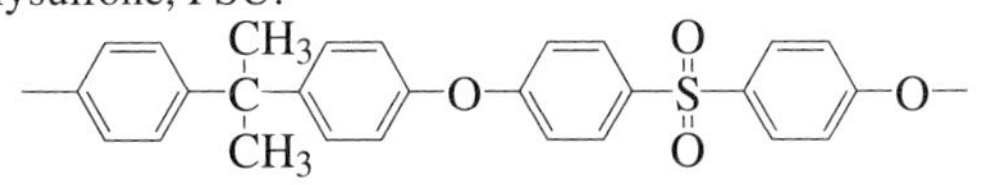

Poly(phenylene sulfide), PPS:

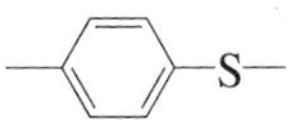

Poly(ether sulfone), PES:

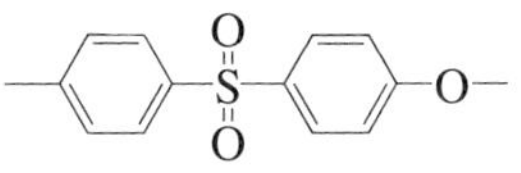

Polyimides and Polyether Ketones
Poly(amide imide), PAI:

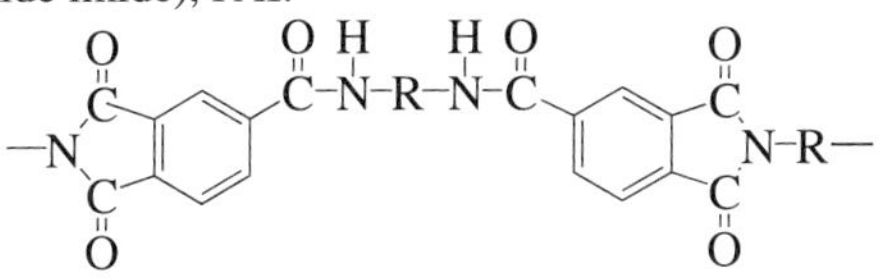

Poly(ether imide), PEI:

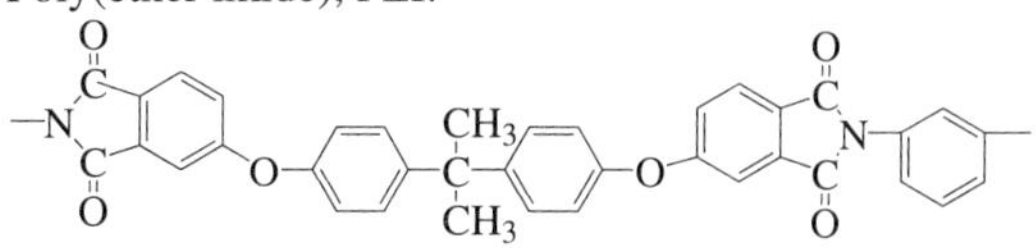

Polyimide, PI:

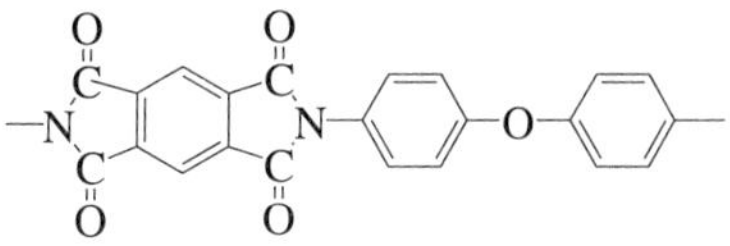

Poly(ether ether ketone), PEEK:

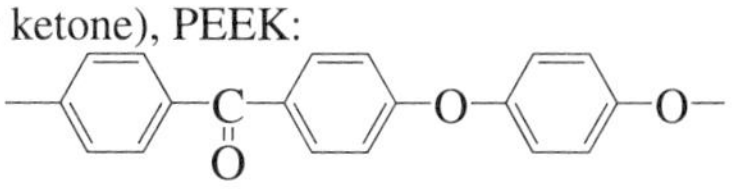

Cellulose

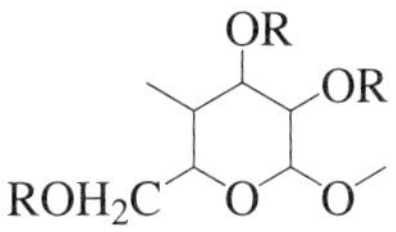

Cellulose acetate, CA: $R = -COCH_3$

Cellulose propionate, CP: $R = -COCH_2CH_3$

Cellulose acetobutyrate, CAB: $R = -COCH_3$ and
$R = -COCH_2CH_2CH_3$

Ethyl cellulose, EC: $R = -CH_2CH_3$

Polyurethanes
Polyurethane, PUR, TPU:
$-CO-NH-(CH_2)_6-NH-CO-O-(CH_2)_4-O-$

Thermosets
Phenol formaldehyde, PF:

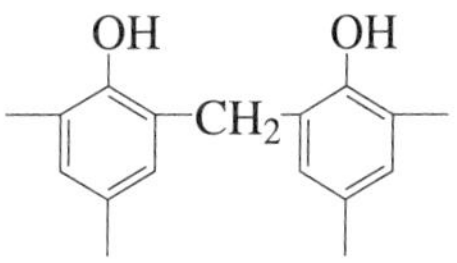

Urea formaldehyde, UF:
$-CH_2-N-$
$C=O$
$-N-CH_2-$

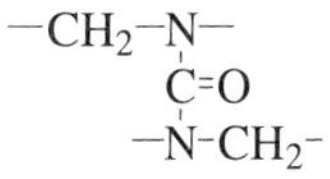

Melamine formaldehyde, MF:

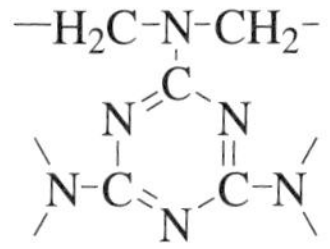

3.3.2 Abbreviations

The following abbreviations are used in this chapter. The abbreviations are in accordance with international rules.

am	amorphous
C	chlorinated
co	copolymer
cr	crystalline
DOP	dioctyl phthalate
HI	high impact (modifier)
iso	isotactic
long	longitudinal
LCP	liquid crystal polymer
mu	monomer unit
pcr	partially crystalline
syn	syndiotactic
str	strained

THF	tetrahydrofuran
trans	transverse
DSC	differential scanning calorimetry
DTA	differential thermal analysis
CFa	carbon fiber content = a mass%, e.g. PS-CF20 means polystyrene with 20% carbon fiber
GBa	content of glass beads, spheres, or balls = a mass%, e.g. PS-GB20 means polystyrene with 20% glass beads
GFa	glass fiber content = a mass%, e.g. PS-GF20 means polystyrene with 20% glass fiber
MeFa	metal fiber content = a mass%

Abbreviated notations for polymer names

ABS	Poly(acrylonitrile-co-butadiene-co-styrene)		PEI	Poly(ether imide)
ASA	Poly(acrylonitrile-co-styrene-co-acrylester)		PES	Poly(ether sulfone)
CA	Cellulose acetate		PET	Poly(ethylene terephthalate)
CAB	Cellulose acetobutyrate		PF	Phenol formaldehyde
COC	Cycloolefine copolymer		PI	Polyimide
CP	Cellulosepropionate		PIB	Poly(isobutylene)
DAP	Diallylphthalat		PMMA	Poly(methyl methacrylate)
EAA	Poly(ethylene-co-acrylic acid)		PMP	Poly(4-methylpenten-1)
EC	Ethylcellulose		POM	Poly(oxymethylene)
ECTFE	Poly(ethylene-co-chlorotrifluoroethylene)		PP	Polypropylene
EPDM	Ethylene/propylene/diene-rubber		PPE	Poly(phenylene ether)
EIM	Polyethylene ionomer		PPS	Poly(phenylene sulfide)
EP	Epoxide; epoxy		PTFE	Poly(tetrafluoro ethylene)
ETFE	Poly(ethylene-co-tetrafluoroethylene)		PVC-P	Poly(vinyl chloride), plasticized with DOP
EVA	Poly(ethylene-co-vinylacetate)		PVC-U	Poly(vinyl chloride), unplasticized
FEP	Poly(tetrafluoroethylene-co-hexafluoropropylene)		PVK	Poly(vinyl carbazole)
HDPE	High density polyethylene		PS	Polystyrene
LDPE	Low density polyethylene		PSU	Polysulfone
LLDPE	Linear low density polyethylene		PUR	Polyurethane
MDPE	Medium density polyethylene		SAN	Poly(styrene-co-acrylonitrile)
MF	Melamine formaldehyde		SB	Poly(styrene-co-butadiene)

PA	Polyamide	SI	Silicone resin
PAI	Poly(amide imide)	TPU	Thermoplastic polyurethane elastomer
PB	Polybutene-1	UF	Urea formaldehyde
PBT	Poly(butylene terephthalate)	UHMWPE	Ultra high molecular weight polyethylene
PC	Polycarbonate	UP	Unsaturated polyester
PCTFE	Poly(trifluorochloroethylene)	VC/A	Poly(vinyl chloride-co-acrylate)
PE	Polyethylene	VF	Vulcanized fiber
PEEK	Poly(ether ether ketone)		

3.3.3 Tables and Figures

The following tables and diagrams contain physical and physicochemical properties of common polymers, copolymers, and polymer blends. The materials are arranged according to increasing number of functional groups, i.e. polyolefines, vinyl polymers, fluoropolymers, polyacrylics, polyacetals, polyamides, polyesters, and polymers with special functional groups [3.2–16].

3.3.3.1 Polyolefines

Polyethylene, HDPE, MDPE. Applications: injection molding for domestic parts and industrial parts; blow molding for containers and sports goods; extrusion for pressure pipes, pipes, electrical insulating material, bags, envelopes, and tissue.

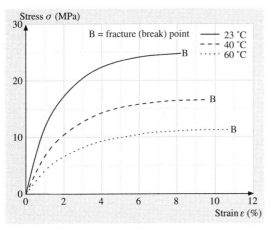

Fig. 3.3-2 Polyethylene, HDPE: stress versus strain

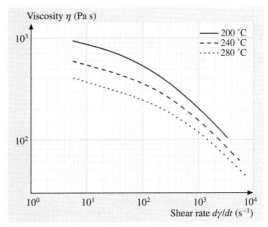

Fig. 3.3-3 Polyethylene, HDPE: viscosity versus shear rate

Table 3.3-1 Polyethylene: high-density, HDPE; medium-density, MDPE; low-density, LDPE; linear low-density, LLDPE; ultrahigh-molecular-weight, UHMWPE

	HDPE	MDPE	LDPE	LLDPE	UHMWPE
Melting temperature T_m (°C)	126–135	120–125	105–118	126	130–135
Enthalpy of fusion ΔH_u (kJ/mol) (mu)	3.9–4.1		3.9–4.1		
Entropy of fusion ΔS_u (J/(K mol)) (mu)	9.6–9.9		9.6–9.9		
Heat capacity c_p (kJ/(kg K)) Temperature coefficient dc_p/dT (kJ/(kg K^2))	2.1–2.7		2.1–2.5		1.7–1.8
Enthalpy of combustion ΔH_c (kJ/g)	−46.4	−46.5	−46.5		
Glass transition temperature T_g (°C)	−110	−110		−110	−110
Vicat softening temperature $T_V 50/50$ (°C)	60–80		45–60		74
Thermal conductivity λ (W/(m K))	0.38–0.51		0.32–0.40		0.41
Density ϱ (g/cm^3)	0.94–0.96	0.925–0.935	0.915–0.92	≈ 0.935	0.93–0.94
Coefficient of thermal expansion α (10^{-5}/K) (linear) (296–328 K)	14–18	18–23	23–25	18–20	15–20
Compressibility κ (10^{-4}/MPa) (cubic)			2.2		
Elastic modulus E (GPa)	0.6–1.4	0.4–0.8	0.2–0.4	0.3–0.7	0.7–0.8
Shear modulus G (GPa)	0.85	0.66	0.16–0.25		
Poisson's ratio μ					
Stress at yield σ_y (MPa)	18–30	11–18	8–10	20–30	≈ 22
Stress at 50% strain σ_{50} (MPa)					
Strain at yield ε_y (%)	8–12	10–15	≈ 20	≈ 15	≈ 15
Stress at fracture σ_b (MPa)	18–35		8–23		
Strain at fracture ε_b (%)	100–1000		300–1000		
Impact strength (Charpy) (kJ/m^2)			13–25		
Notched impact strength (Charpy) (kJ/m^2)			3–5		
Sound velocity v_s (m/s) (longitudinal)	2430		2400		
Sound velocity v_s (m/s) (transverse)	950		1150		
Shore hardness D	58–63	45–60	45–51	38–60	62
Volume resistivity ρ_e (Ω m) > 10^{15}	> 10^{15}	> 10^{15}	> 10^{15}	> 10^{15}	
Surface resistivity σ_e (Ω)	> 10^{14}	> 10^{14}	> 10^{14}	> 10^{14}	> 10^{14}
Electric strength E_B (kV/mm)	30–40	30–40	30–40	30–40	30–40
Relative permittivity ε_r (100 Hz)	2.4	2.3	2.3	2.3	2–2.4
Dielectric loss $\tan\delta$ (10^{-4}) (100 Hz)	1–2	2	2–2.4	2	2
Refractive index n_D (589 nm) Temperature coefficient dn_D/dT (10^{-4}/K)	1.53		1.51–1.42		
Steam permeation (g/(m^2 d))	0.9 (40 μm)		1 (100 μm)		
Gas permeation (cm^3/(m^2 d bar)), 23 °C, 100 μm	700 (N$_2$) 1800 (O$_2$) 10 000 (CO$_2$) 1100 (air)		700 (N$_2$) 2000 (O$_2$) 10 000 (CO$_2$) 1100 (air)		
Melt viscosity–molar-mass relation					
Viscosity–molar-mass relation	$[\eta] = 62 \times 10^{-3} M^{0.70}$ (decalin, 135 °C) $[\eta] = 51 \times 10^{-3} M^{0.725}$ (tetralin, 130 °C)				

Table 3.3-2 Polyethylene, HDPE: heat capacity, thermal conductivity, and coefficient of thermal expansion

Temperature T (°C)	−200	−150	−100	−50	0	20	50	100	150
Heat capacity c_p (kJ/(kg K))	0.55	0.84	1.10	1.34	1.64		2.05	2.86	
Thermal conductivity λ (W/(m K))		0.62	0.56	0.50	0.44		0.38	0.32	0.25
Coefficient of thermal expansion α (10^{-5}/K) (linear)	4.5	6.8	9.5	12.4		16.9	33.0	69.0	

Table 3.3-3 Polyethylene, LDPE: heat capacity and thermal conductivity

Temperature T (°C)	−200	−150	−100	−50	0	20	50	100	150
Heat capacity c_p (kJ/(kg K))	0.55	0.84	1.10	1.43	1.90		2.73		
Thermal conductivity λ (W/(m K))		0.36	0.38	0.38	0.35		0.31	0.24	0.25

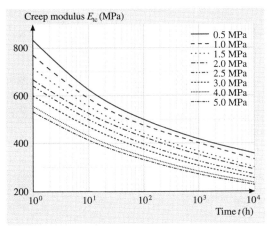

Fig. 3.3-4 Polyethylene, HDPE: creep modulus versus time, at 23 °C

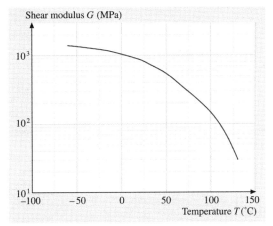

Fig. 3.3-5 Polyethylene, HDPE: shear modulus versus temperature

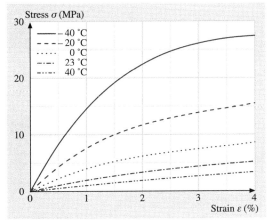

Fig. 3.3-6 Polyethylene, LDPE: stress versus strain

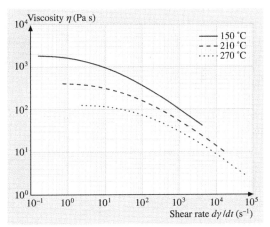

Fig. 3.3-7 Polyethylene, LDPE: viscosity versus shear rate

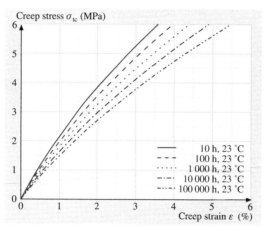

Fig. 3.3-8 Polyethylene, UHMWPE: isochronous stress versus strain

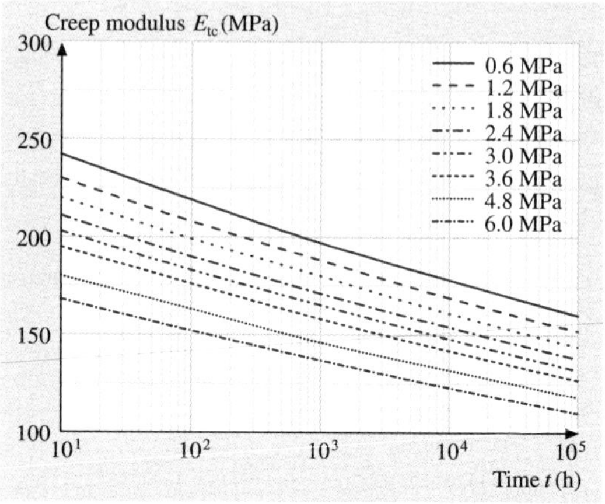

Fig. 3.3-9 Polyethylene, UHMWPE: creep modulus versus time

Polyethylene, LDPE. Applications: all kinds of sheeting, bags, insulating material, hollow bodies, bottles, injection-molded parts.

Polyethylene, LLDPE. Applications: foils, bags, waste bags, injection-molded parts, rotatory-molded parts.

Polyethylene UHMWPE, Applications: mechanical engineering parts, food wrapping, commercial packaging, textile industry parts, electrical engineering parts, paper industry parts, low temperature materials, medical parts, chemical engineering parts, electroplating.

Poly(ethylene-co-vinyl acetate), EVA. Applications: films, deep-freeze packaging, laminates, fancy leather, food packaging, closures, ice trays, bags, gloves, fittings, pads, gaskets, plugs, toys.

Polyethylene ionomer, EIM. Applications: transparent tubes for water and liquid foods, transparent films, bottles, transparent coatings, adhesion promoters.

Cycloolefine copolymer, COC. Applications: precision optics, optical storage media, lenses, medical and labware applications.

Poly(ethylene-co-acrylic acid), EAA. Applications: packaging foils, sealing layers, tubing.

Polypropylene, PP. Applications: injection molding for domestic parts, car parts, electric appliances, packing materials, and pharmaceutical parts; blow

Table 3.3-4 Poly(ethylene-co-vinyl acetate), EVA; polyethylene ionomer, EIM; cycloolefine copolymer, COC (poly-(ethylene-co-norbornene)); poly(ethylene-co-acrylic acid), EAA

	EVA	EIM	COC	EAA
Melting temperature T_m (°C)	90–110	95–110		92–103
Enthalpy of fusion ΔH_u (kJ/mol) (mu)				
Entropy of fusion ΔS_u (J/(K mol)) (mu)				
Heat capacity c_p (kJ/(kg K))		2.2		
Temperature coefficient dc_p/dT (kJ/(kg K²))				
Enthalpy of combustion ΔH_c (kJ/g)				
Glass transition temperature T_g (°C)	66		80–180	
Vicat softening temperature T_V 50/50 (°C)	63–96 VST/A50			
Thermal conductivity λ (W/(m K))	0.28–0.35	0.24	0.16	
Density ϱ (g/cm³)	0.93–0.94	0.94–0.95	1.02	0.925–0.935
Coefficient of thermal expansion α (10^{-5}/K) (linear)	≈ 25	10–15	6–7	≈ 20

Table 3.3-4 Poly(ethylene-co-vinyl acetate), EVA; polyethylene ionomer, EIM; ..., cont.

	EVA	EIM	COC	EAA
Compressibility κ (10^{-4}/MPa) (cubic)				
Elastic modulus E (GPa)	0.03–0.1	0.15–0.2	2.6–3.2	0.04–0.13
Shear modulus G (GPa)	0.04–0.14			
Poisson's ratio μ				
Stress at yield σ_y (MPa)	5–8	7–8	66	4–7
Strain at yield ε_y (%)		> 20	3.5–10	> 20
Stress at 50% strain σ_{50} (MPa)	4–9			
Stress at fracture σ_b (MPa)	16–23	21–35	66	
Strain at fracture ε_b (%)	700	250–500	3–10	
Impact strength (Charpy) (kJ/m^2)			13–20	
Notched impact strength (Charpy) (kJ/m^2)			1.7–2.6	
Sound velocity v_s (m/s) (longitudinal)				
Sound velocity v_s (m/s) (transverse)				
Shore hardness D	34–44			
Volume resistivity ρ_e (Ω m)	> 10^{14}	> 10^{15}	> 10^{13}	> 10^{14}
Surface resistivity σ_e (Ω)	> 10^{13}	> 10^{13}	10^{14}	> 10^{13}
Electric strength E_B (kV/mm)	30–35	40		30–40
Relative permittivity ε_r (100 Hz)	2.5–3	≈ 2.4	2.35	2.5–3
Dielectric loss $\tan\delta$ (10^{-4}) (100 Hz)	20–40	≈ 30	0.2	30–130
Refractive index n_D (589 nm)		1.51	1.53	
Temperature coefficient dn_D/dT (10^{-4}/K)				
Steam permeation (g/(m^2 d)) (25 µm)		25	1–1.8	
Gas permeation (cm^3/(m^2 d bar)) (25 µm)		9300 (O$_2$)		
Melt viscosity–molar-mass relation				
Viscosity–molar-mass relation				

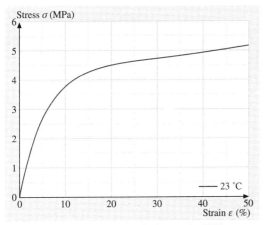

Fig. 3.3-10 Poly(ethylene-co-vinylacetate), EVA: stress versus strain

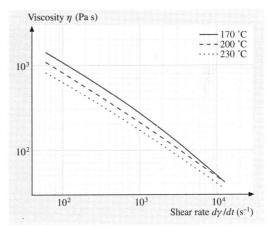

Fig. 3.3-11 Poly(ethylene-co-vinylacetate), EVA: viscosity versus shear rate

Table 3.3-5 Polypropylene, PP; polybutene-1, PB; polyisobutylene, PIB; poly(4-methylpentene-1), PMP

	PP	PB	PIB	PMP
Melting temperature T_m (°C)	160–170	126		230–240
Enthalpy of fusion ΔH_u (kJ/mol) (mu)	7–10			
Entropy of fusion ΔS_u (J/(K mol)) (mu)	15–20			
Heat capacity c_p (kJ/(kg K))	1.68			
Temperature coefficient dc_p/dT (kJ/(kg K^2))				
Enthalpy of combustion ΔH_c (kJ/g)	−44		−45	
Glass transition temperature T_g (°C)	0 to −10	78	−70	
Vicat softening temperature T_V 50/50 (°C)	60–102	108–113		
Thermal conductivity λ (W/(m K))	0.22	0.22	0.12–0.20	0.17
Density ϱ (g/cm^3)	0.90–0.915	0.905–0.920	0.91–0.93	0.83
Coefficient of thermal expansion α (10^{-5}/K) (linear)	12–15	13	8–12	12
Compressibility κ (10^{-4}/MPa) (cubic)	2.2			
Elastic modulus E (GPa)	1.3–1.8	0.21–0.26		1.1–2.0
Shear modulus G (GPa)				
Poisson's ratio μ				
Stress at yield σ_y (MPa)	25–40	16–24		
Strain at yield ε_y (%)	8–18	24		
Stress at 50% elongation σ_{50} (MPa)				
Stress at fracture σ_b (MPa)		30–38	2–6	25–28
Strain at fracture ε_b (%)	> 50	300–380	> 1000	10–50
Impact strength (Charpy) (kJ/m^2)	60–140			
Notched impact strength (Charpy) (kJ/m^2)	3–10			
Sound velocity v_s (m/s) (longitudinal)	2650		1950	2180
Sound velocity v_s (m/s) (transverse)	1300			1080
Shore hardness D	69–77			
Volume resistivity ρ_e (Ω m)	> 10^{14}	> 10^{14}	> 10^{13}	> 10^{14}
Surface resistivity σ_e (Ω)	> 10^{13}			
Electric strength E_B (kV/mm)	30–70	18–40	23–25	
Relative permittivity ε_r (100 Hz)	2.3	2.5	2.3	2.1
Dielectric loss $\tan\delta$ (10^{-4}) (100 Hz)	2.5	20–50	< 4	15
Refractive index n_D (589 nm)	1.50			1.46
Temperature coefficient dn_D/dT (10^{-4}/K)				
Steam permeation (g/(m^2 d)) (40 μm)	2.1			
Gas permeation (cm^3/(m^2 d bar)) (40 μm)	430 (N$_2$) 1900 (O$_2$) 6100 (CO$_2$) 700 (air)			
Melt viscosity–molar-mass relation				
Viscosity–molar-mass relation	[η] = 20 × 10^{-3} $M^{0.67}$ (PIB, toluene, 30 °C)			

Table 3.3-6 Polypropylene, PP: heat capacity, thermal conductivity, and coefficient of thermal expansion

Temperature T (°C)	−200	−150	−100	−50	0	20	50	100	150
Heat capacity c_p (kJ/(kg K))	0.46	0.77	1.03	1.28	1.57	1.68	1.92	2.35	
Thermal conductivity λ (W/(m K))	2.3	0.17	0.19	0.21	0.22	0.22	0.22	0.20	
Coefficient of thermal expansion α (10^{-5}/K) (linear)		5.8	6.9	7.6	19.1	19.4	14.3	22.6	29.4

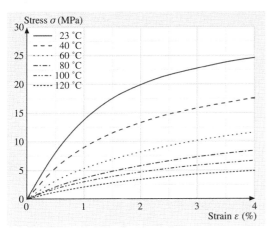

Fig. 3.3-12 Polypropylene, PP: stress versus strain

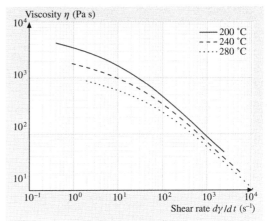

Fig. 3.3-13 Polypropylene, PP: viscosity versus shear rate

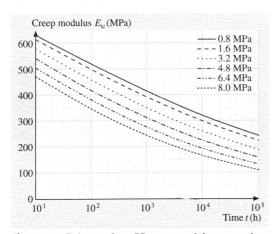

Fig. 3.3-14 Polypropylene, PP: creep modulus versus time, at 23 °C

molding for cases, boxes, and bags; pressure molding for blocks and boards; tissue coating; extrusion for pressure pipes, pipes, semiproducts, flexible tubing, textile fabrics, fibers, sheets; foamed polymers for sports goods, boxes, cases, handles, and domestic parts.

Polybutene-1, PB. Applications: tubing, pipes, fittings, hollow bodies, cables, foils, packing.

Polyisobutylene, PIB. Applications: adhesives, sealing compounds, electrical insulating oils, viscosity improvers.

Poly(4-methylpentene-1), PMP. Applications: glasses, light fittings, molded parts, foils, packing, rings, tubes, fittings, cable insulation.

3.3.3.2 Vinyl Polymers

Polystyrene, PS. Applications: packaging, expanded films, toys, artificial wood, food holders, housings, plates, cases, boxes, glossy sheets, containers, medical articles, panels.

Poly(styrene-co-butadiene), SB. Applications: flame-retardant articles, housings.

Poly(styrene-co-acrylnitrile), SAN. Applications: food holders, cases, shelves, covers, automotive and electrical parts, sheets, profiles, containers, plastic windows, doors, cosmetic items, medical items, pharmaceutical items, fittings, components, musical items, toys, displays, lighting, food compartments, cassettes, tableware, engineering parts, housings.

Table 3.3-7 Polystyrene, PS; poly(styrene-co-butadiene), SB; poly(styrene-co-acrylonitrile), SAN

	PS	SB	SAN
Melting temperature T_m (°C)	> 243.2 (iso), > 287.5 (syn)		
Enthalpy of fusion ΔH_u (kJ/mol) (mu)	8.682 (iso), 8.577 (syn)		
Entropy of fusion ΔS_u (J/(K mol)) (mu)	16.8 (iso), 15.3 (syn)		
Heat capacity c_p (kJ/(kg K))	1.2–1.4	1.3	1.3
Temperature coefficient dc_p/dT (kJ/(kg K^2))	4.04×10^{-3} (323 K)		
Enthalpy of combustion ΔH_c (kJ/g)	−41.6	−36 to −38	−36 to −38
Glass transition temperature T_g (°C)	95–100	90–95	110
Vicat softening temperature T_V 50/50 (°C)	80–100	80–110	105–120
Thermal conductivity λ (W/(m K))	0.16	0.18	0.18
Density ϱ (g/cm^3)	1.05	1.05–1.06	1.07–1.08
Coefficient of thermal expansion α (10^{-5}/K) (linear)	6–8	8–10	7–8
Compressibility κ (10^{-4}/MPa) (cubic)	2.2		
Elastic modulus E (GPa)	3.1–3.3	2.0–2.8	3.5–3.9
Shear modulus G (GPa)	1.2	0.6	1.5
Poisson's ratio μ	0.38		
Stress at yield σ_y (MPa)	50	25–45	
Strain at yield ε_y (%)		1.1–2.5	
Stress at 50% elongation σ_{50} (MPa)			
Stress at fracture σ_b (MPa)	30–55	26–38	65–85
Strain at fracture ε_b (%)	1.5–3	25–60	2.5–5
Impact strength (Charpy) (kJ/m^2)	13–25	50–105	
Notched impact strength (Charpy) (kJ/m^2)	3–5	5–10	2–4
Sound velocity v_s (m/s) (longitudinal)	2400		
Sound velocity v_s (m/s) (transverse)	1150		
Shore hardness D	78		
Volume resistivity ρ_e (Ω m)	> 10^{14}	> 10^{14}	10^{14}
Surface resistivity σ_e (Ω)	> 10^{14}	> 10^{13}	10^{14}
Electric strength E_B (kV/mm)	30–70	45–65	30–60
Relative permittivity ε_r (100 Hz)	2.4–2.5 (am); 2.61 (cr)	2.4–2.6	2.8–3
Dielectric loss $\tan\delta$ (10^{-4}) (100 Hz)	1–2	1–3	40–50
Refractive index n_D (589 nm)	1.58–1.59		1.57
Temperature coefficient dn_D/dT (10^{-4}/K)	−1.42		
Steam permeation (g/(m^2 d)) (100 μm)	12		
Gas permeation (cm^3/(m^2 d bar)) (100 μm)	2500 (N$_2$) 1000 (O$_2$) 5200 (CO$_2$)		
Melt viscosity–molar-mass relation	$\eta = 13.04 M^{3.4}$ (217 °C)		
Viscosity–molar-mass relation	$[\eta] = 11.3 \times 10^{-3} M^{0.73}$ (PS, toluene, 25 °C)		

Table 3.3-8 Polystyrene, PS: heat capacity, thermal conductivity, and coefficient of thermal expansion

Temperature T (°C)	−200	−150	−100	−50	0	20	50	100	150	200
Heat capacity c_p (kJ/(kg K))				0.89	1.09		1.34	1.68	2.03	2.19
Thermal conductivity λ (W/(m K))		0.13	0.14	0.14	0.16	0.16	0.16	0.16	0.16	
Coefficient of thermal expansion α (10^{-5}/K) (linear)	3.9	5.1	6.1	6.7		7.1	10.0	17.6	18.0	17.4

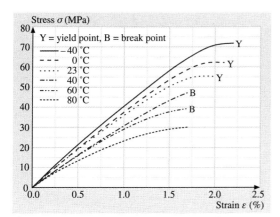

Fig. 3.3-15 Polystyrene, PS: stress versus strain

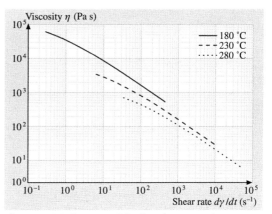

Fig. 3.3-16 Polystyrene, PS: viscosity versus shear rate

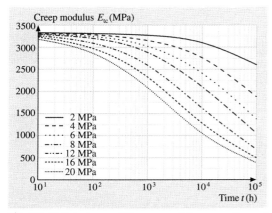

Fig. 3.3-17 Polystyrene, PS: creep modulus versus time, at 23 °C

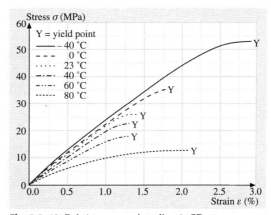

Fig. 3.3-18 Poly(styrene-co-butadiene), SB: stress versus strain

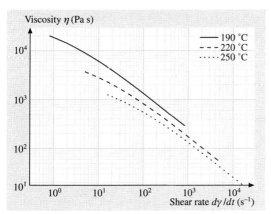

Fig. 3.3-19 Poly(styrene-co-butadiene), SB: viscosity versus shear rate

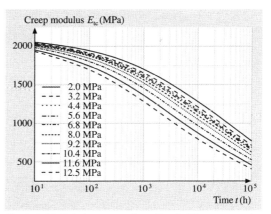

Fig. 3.3-20 Poly(styrene-co-butadiene), SB: creep modulus versus time

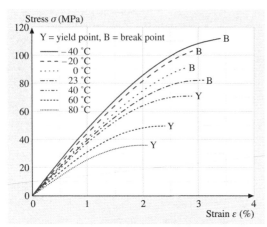

Fig. 3.3-21 Poly(styrene-co-acrylonitrile), SAN: stress versus strain

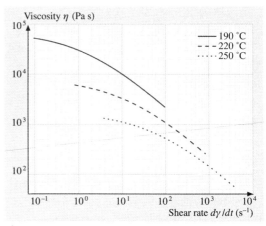

Fig. 3.3-22 Poly(styrene-co-acrylonitrile), SAN: viscosity versus shear rate

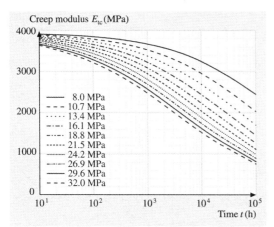

Fig. 3.3-23 Poly(styrene-co-acrylonitrile), SAN: creep modulus versus time

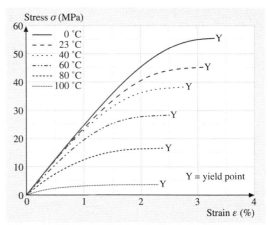

Fig. 3.3-24 Poly(acrylonitrile-co-butadiene-co-styrene), ABS: stress versus strain

Poly(vinyl carbazole), PVK. Applications: electrical insulating materials.

Poly(acrylonitrile-co-butadiene-co-styrene), ABS. Applications: housings, boxes, cases, tools, office equipment, helmets, pipes, fittings, frames, toys, sports equipment, housewares, packaging, panels, covers, automotive interior parts, food packaging, furniture, automobile parts.

Poly(acrylonitrile-co-styrene-co-acrylester), ASA. Applications: outdoor applications, housings, covers, sports equipment, fittings, garden equipment, antennas.

Poly(vinyl chloride), PVC-U. Applications: extruded profiles, sealing joints, jackets, furniture, claddings, roller shutters, fences, barriers, fittings, injection-molded articles, sheets, films, plates, bottles, coated fabrics, layers, toys, bumpers, buoys, car parts, cards, holders, sleeves, pipes, boxes, inks, lacquers, adhesives, foams.

Poly(vinyl chloride), PVC-P1, PVC-P2. Applications: flexible tubes, technical items, compounds for insulation and jacketing, joints, seals, pipes, profiles, soling, tubes, floor coverings, wall coverings, car undersealing, mastics, foams, coatings, cap closures,

Table 3.3-9 Poly(vinyl carbazole), PVK; poly(acrylonitrile-co-butadiene-co-styrene), ABS; poly(acrylonitrile-co-styrene-co-acrylester), ASA

	PVK	ABS	ASA
Melting temperature T_m (°C)			
Enthalpy of fusion ΔH_u (kJ/mol) (mu)			
Entropy of fusion ΔS_u (J/(K mol)) (mu)			
Heat capacity c_p (kJ/(kg K))		1.3	1.3
Temperature coefficient dc_p/dT (kJ/(kg K^2))			
Enthalpy of combustion ΔH_c (kJ/g)		−35	
Glass transition temperature T_g (°C)	173	80–110	100
Vicat softening temperature T_V 50/50 (°C)		95–105	90–102
Thermal conductivity λ (W/(m K))	0.29	0.18	0.18
Density ϱ (g/cm^3)		1.03–1.07	1.07
Coefficient of thermal expansion α (10^{-5}/K) (linear)	1.19	8.5–10	9.5
Compressibility κ (1/MPa) (cubic)			
Elastic modulus E (GPa)	3.5	2.2–3.0	2.3–2.9
Shear modulus G (GPa)		0.7–0.9	0.7–0.9
Poisson's ratio μ			
Stress at yield σ_y (MPa)		45–65	40–55
Strain at yield ε_y (%)		2.5–3	3.1–4.3
Stress at 50% elongation σ_{50} (MPa)			
Stress at fracture σ_b (MPa)	20–30	15–30	
Strain at fracture ε_b (%)		55–80	
Impact strength (Charpy) (kJ/m^2)		40–1000	105–118
Notched impact strength (Charpy) (kJ/m^2)			
Sound velocity v_s (m/s) (longitudinal)		2040–2160	
Sound velocity v_s (m/s) (transverse)		830–930	
Shore hardness D			
Volume resistivity ρ_e (Ω m)	$> 10^{14}$	10^{12}–10^{13}	10^{12}–10^{14}
Surface resistivity σ_e (Ω)	10^{14}	$> 10^{13}$	10^{13}
Electric strength E_B (kV/mm)		30–40	
Relative permittivity ε_r (100 Hz)		2.8–3.1	3.4–4
Dielectric loss tan δ (10^{-4}) (100 Hz)	6–10	90–160	90–100
Refractive index n_D (589 nm)		1.52	
Temperature coefficient dn_D/dT (1/K)			
Steam permeation (g/(m^2 d)) (100 μm)		27–33	30–35
Gas permeation (cm^3/(m^2 d bar)) (100 μm)		100–200 (N$_2$)	60–70 (N$_2$)
		400–900 (O$_2$)	150–180 (O$_2$)
			6000–8000 (CO$_2$)
Melt viscosity–molar-mass relation			
Viscosity–molar-mass relation			

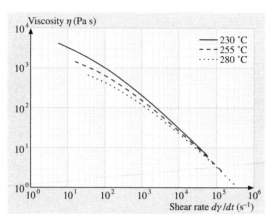

Fig. 3.3-25 Poly(acrylonitrile-co-butadiene-co-styrene), ABS: viscosity versus shear rate

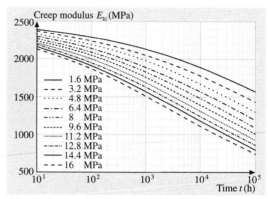

Fig. 3.3-26 Poly(acrylonitrile-co-butadiene-co-styrene), ABS: creep modulus versus time

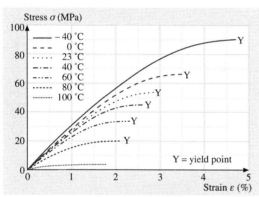

Fig. 3.3-27 Poly(acrylonitrile-co-styrene-co-acrylester), ASA: stress versus strain

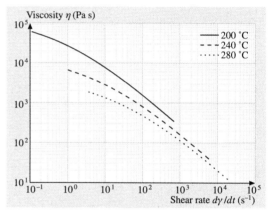

Fig. 3.3-28 Poly(acrylonitrile-co-styrene-co-acrylester), ASA: viscosity versus shear rate

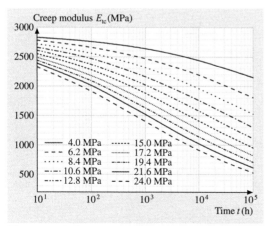

Fig. 3.3-29 Poly(acrylonitrile-co-styrene-co-acrylester), ASA: creep modulus versus time

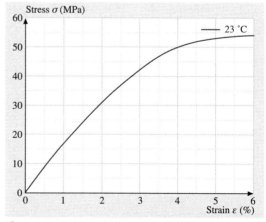

Fig. 3.3-30 Poly(vinyl chloride), PVC-U: stress versus strain

Table 3.3-10 Poly(vinyl chloride): unplasticized, PVC-U; plasticized (75/25), PVC-P1; plasticized (60/40), PVC-P2

	PVC-U	PVC-P1	PVC-P2
Melting temperature T_m (°C)	175		
Enthalpy of fusion ΔH_u (kJ/mol) (mu)	3.28		
Entropy of fusion ΔS_u (J/(K mol)) (mu)			
Heat capacity c_p (kJ/(kg K))	0.85–0.9	0.9–1.8	0.9–1.8
Temperature coefficient dc_p/dT (kJ/(kg K^2))			
Enthalpy of combustion ΔH_c (kJ/g)	−18 to −19		
Glass transition temperature T_g (°C)	85	≈ 80	≈ 80
Vicat softening temperature T_V 50/50 (°C)	63–82	≈ 42	
Thermal conductivity λ (W/(m K))	0.16	0.15	0.15
Density ϱ (g/cm^3)	1.38–1.4	1.24–1.28	1.15–1.20
Coefficient of thermal expansion α (10^{-5}/K) (linear)	7–8	18–22	23–25
Compressibility κ (1/MPa) (cubic)			
Elastic modulus E (GPa)	2.7–3.0	2.9–37	
Shear modulus G (GPa)	0.12	0.07	
Poisson's ratio μ			
Stress at yield σ_y (MPa)	50–60		
Strain at yield ε_y (%)	4–6		
Stress at 50% elongation σ_{50} (MPa)			
Stress at fracture σ_b (MPa)	40–80	10–25	10–25
Strain at fracture ε_b (%)	10–50	170–400	170–400
Impact strength (Charpy) (kJ/m^2)	13–25		
Notched impact strength (Charpy) (kJ/m^2)	3–5		
Sound velocity v_s (m/s) (longitudinal)	2330	2126	
Sound velocity v_s (m/s) (transverse)	1070		
Shore hardness D	74–94		
Volume resistivity ρ_e (Ω m)	> 10^{13}	10^{12}	10^{11}
Surface resistivity σ_e (Ω)	10^{14}	10^{11}	10^{10}
Electric strength E_B (kV/mm)	20–40	30–35	≈ 25
Relative permittivity ε_r (100 Hz)	3.5	4–5	6–7
Dielectric loss tan δ (10^{-4}) (100 Hz)	110–140	0.05–0.07	0.08–0.1
Refractive index n_D (589 nm)	1.52–1.54		
Temperature coefficient dn_D/dT (1/K)			
Steam permeation (g/(m^2 d)) (100 μm)	2.5	20	20
Gas permeation (cm^3/(m^2 d bar)) (100 μm)	2.7–3.8 (N$_2$)	350 (N$_2$)	350 (N$_2$)
	33–45 (O$_2$)	1500 (O$_2$)	1500 (O$_2$)
	120–160 (CO$_2$)	8500 (CO$_2$)	8500 (CO$_2$)
	28 (air)	550 (air)	550 (air)
Melt viscosity–molar-mass relation			
Viscosity–molar-mass relation	$[\eta] = 14.5 \times 10^{-3} M^{0.851}$ (PVC-U, THF, 25 °C)		

Table 3.3-11 Poly(vinyl chloride), PVC-U: heat capacity and thermal conductivity

Temperature T (°C)	−150	−100	−50	0	20	50	100
Heat capacity c_p (kJ/(kg K))	0.47	0.61	0.75	0.92		1.04	1.53
Thermal conductivity λ (W/(m K))	0.13	0.15	0.15	0.16	0.16	0.17	0.17

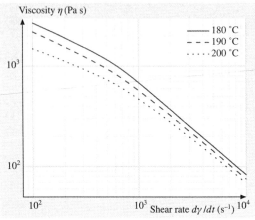

Fig. 3.3-31 Poly(vinyl chloride), PVC-U: viscosity versus shear rate

protective clothes, cable insulation, fittings, leads, films, artificial leathers.

3.3.3.3 Fluoropolymers

Polytetrafluoroethylene, PTFE. Applications: molded articles, foils, flexible tubing, coatings, jackets, sealings, bellows, flasks, machine parts, semifinished goods, printed networks.

Polychlorotrifluoroethylene, PCTFE. Applications: fittings, flexible tubing, membranes, printed networks, bobbins, insulating foils, packings.

Poly(tetrafluoroethylene-co-hexafluoropropylene), FEP. Applications: cable insulation, coatings, coverings, printed networks, injection-molded parts, packaging foils, impregnations, heat-sealing adhesives.

Table 3.3-12 Polytetrafluoroethylene, PTFE; polychlorotrifluoroethylene, PCTFE; poly(tetrafluoroethylene-co-hexafluoro-propylene), FEP; poly(ethylene-co-tetrafluoroethylene), ETFE; poly(ethylene-co-chlorotrifluoroethylene), ECTFE

	PTFE	PCTFE	FEP	ETFE	ECTFE
Melting temperature T_m (°C)	325–335	210–215	255–285	265–270	240
Enthalpy of fusion ΔH_u (kJ/kg)		1.2	24.3		
Entropy of fusion ΔS_u (J/(kg K))					
Heat capacity c_p (kJ/(kg K))	1.0	0.9	1.12	0.9	
Temperature coefficient dc_p/dT (kJ/(kg K^2))					
Enthalpy of combustion ΔH_c (kJ/g)	−5.1				
Glass transition temperature T_g (°C)	127				
Vicat softening temperature $T_V 50/50$ (°C)	110			134	
Thermal conductivity λ (W/(m K))	0.25	0.22	0.25	0.23	
Density ϱ (g/cm^3)	2.13–2.23	2.07–2.12	2.12–2.18	1.67–1.75	1.68–1.70
Coefficient of thermal expansion α (10^{-5}/K) (linear)	11–18	6–7	8–12	7–10	7–8
Compressibility κ (1/MPa) (cubic)					
Elastic modulus E (GPa)	0.40–0.75	1.30–1.50	0.40–0.70	0.8–1.1	1.4–1.7
Shear modulus G (GPa)					
Poisson's ratio μ					
Stress at yield σ_y (MPa)	11.7			25–35	
Strain at yield ε_y (%)				15–20	

Table 3.3-12 Polytetrafluoroethylene, PTFE; polychlorotrifluoroethylene, PCTFE; ..., cont.

	PTFE	PCTFE	FEP	ETFE	ECTFE
Stress at 50% elongation σ_{50} (MPa)					
Stress at fracture σ_b (MPa)	25–36	32–40	22–28	35–54	
Strain at fracture ε_b (%)	350–550	120–175	250–330	400–500	
Impact strength (Charpy) (kJ/m^2)					
Notched impact strength (Charpy) (kJ/m^2)					
Sound velocity v_s (m/s) (longitudinal)	1410				
Sound velocity v_s (m/s) (transverse)	730				
Shore hardness D	50–60	78	55–58	67–75	
Volume resistivity ρ_e (Ω m)	$>10^{16}$	$>10^{16}$	$>10^{16}$	$>10^{14}$	$>10^{13}$
Surface resistivity σ_e (Ω)	$>10^{16}$	$>10^{16}$	$>10^{16}$	$>10^{14}$	10^{12}
Electric strength E_B (kV/mm)	48	55	55	40	
Relative permittivity ε_r (100 Hz)	2.1	2.5–2.7	2.1	2.6	2.3–2.6
Dielectric loss $\tan\delta$ (10^{-4}) (100 Hz)	0.5–0.7	90–140	0.5–0.7	5–6	10–15
Refractive index n_D (589 nm)	1.35	1.43	1.344	1.403	
Temperature coefficient dn_D/dT (10^{-5}/K)					
Steam permeation (g/(m^2 d))	0.03 (300 µm)	0.4–0.9 (25 µm)		0.6 (25 µm)	9 (25 µm)
Gas permeation (cm^3/(m^2 d bar))	60–80 (N$_2$)	39 (N$_2$)		470 (N$_2$)	150 (N$_2$)
	160–250 (O$_2$)	110–230 (O$_2$)		1560 (O$_2$)	39 (O$_2$)
	450–700 (CO$_2$)	250–620 (CO$_2$)		3800 (CO$_2$)	1700 (CO$_2$)
	80–100 (air)				
Melt viscosity–molar-mass relation					
Viscosity–molar-mass relation					

Table 3.3-13 Polytetrafluoroethylene, PTFE: heat capacity, thermal conductivity, and coefficient of thermal expansion

Temperature T (°C)	−200	−150	−100	−50	0	20	50	100	150	200	250
Heat capacity c_p (kJ/(kg K))				0.8	0.96	1.0	1.06	1.10	1.15	1.23	
Thermal conductivity λ (W/(m K))		0.23	0.24	0.25	0.25	0.25	0.26	0.26	0.26		
Coefficient of thermal expansion α (10^{-5}/K) (linear)	3.4	4.5	7.0	9.5	11.6		11.9	13.1	16.7	22.2	30.5

Poly(ethylene-co-tetrafluoroethylene), ETFE. Applications: gear wheels, pump parts, packaging, laboratory articles, coverings, cable insulation, blown films.

Poly(ethylene-co-chlorotrifluoroethylene), ECTFE. Applications: cable coverings, coverings, molded articles, packaging, printed networks, foils, fibers.

3.3.3.4 Polyacrylics and Polyacetals

Poly(methyl methacrylate), PMMA. Applications: extruded articles, injection-molded parts, injection blow-molded parts, houseware, medical devices, sanitary ware, automotive components, lighting, tubes, sheets, profiles, covers, panels, fiber optics, optical lenses, displays, disks, cards.

Poly(oxymethylene), POM-H. Applications: molded parts, extruded parts, gears, bearings, snap-fits, fuel system components, cable ties, automotive components, seatbelt parts, pillar loops, tubes, panels.

Poly(oxymethylene-co-ethylene), POM-R. Applications: extruded articles, injection-molded articles, gear wheels, bushes, bearings, rollers, guide rails, tubing, films, clips, zippers, boards, pipes.

Table 3.3-14 Poly(methyl methacrylate), PMMA; poly(oxymethylene), POM-H; poly(oxymethylene-co-ethylene), POM-R

	PMMA	POM-H	POM-R
Melting temperature T_m (K)	175	175	164–172
Enthalpy of fusion ΔH_u (kJ/kg)		222	211
Entropy of fusion ΔS_u (J/(kg K))		1.49	1.44
Heat capacity c_p (kJ/(kg K))	1.255	1.46	1.47
Temperature coefficient dc_p/dT (kJ/(kg K^2))			
Enthalpy of combustion ΔH_c (kJ/g)	−26.2	−16.7	−17
Glass transition temperature T_g (°C)	104–105	25	
Vicat softening temperature T_V 50/50 (°C)	85–110	160–170	151–162
Thermal conductivity λ (W/(m K))	0.193	0.25–0.30	0.31
Density ϱ (g/cm^3)	1.17–1.19	1.40–1.42	1.39–1.41
Coefficient of thermal expansion α (10^{-5}/K) (linear)	7–8	11–12	10–11
Compressibility κ (10^{-4}/MPa) (cubic)	2.45	1.5	
Elastic modulus E (GPa)	3.1–3.3	3.0–3.2	2.8–3.2
Shear modulus G (GPa)	1.7	0.7–1.0	
Poisson's ratio μ			
Stress at yield σ_y (MPa)	50–77	60–75	65–73
Strain at yield ε_y (%)		8–25	8–12
Stress at 50% elongation σ_{50} (MPa)			
Stress at fracture σ_b (MPa)	60–75	62–70	59
Strain at fracture ε_b (%)	2–6	25–70	
Impact strength (Charpy) (kJ/m^2)	18–23	180	
Notched impact strength (Charpy) (kJ/m^2)	2	9	6.5
Sound velocity v_s (m/s) (longitudinal)	2690	2440	
Sound velocity v_s (m/s) (transverse)	1340	1000	
Shore hardness D	85	80	
Volume resistivity ρ_e (Ω m)	> 10^{13}	> 10^{13}	> 10^{13}
Surface resistivity σ_e (Ω)	> 10^{13}	> 10^{14}	> 10^{13}
Electric strength E_B (kV/mm)	30	25–35	35
Relative permittivity ε_r (100 Hz)	3.5–3.8	3.5–3.8	3.6–4
Dielectric loss tan δ (10^{-4}) (100 Hz)	500–600	30–50	30–50
Refractive index n_D (589 nm)	1.492	1.49	
Temperature coefficient dn_D/dT (10^{-4}/K)	−1.2		
Steam permeation (g/(m^2 d)) (80 μm)	12	12	
Gas permeation (cm^3/(m^2 d bar))	2500 (N$_2$)	5 (N$_2$)	
	1000 (O$_2$)	24 (O$_2$)	
	5200 (CO$_2$)	470 (CO$_2$)	
		8 (air)	
Melt viscosity–molar-mass relation			
Viscosity–molar-mass relation	$[\eta] = 10.4 \times 10^{-3} M^{0.70}$ (PMMA, THF, 25 °C)		
	$[\eta] = 11.3 \times 10^{-3} M^{0.76}$ (POM-H, Phenol, 90 °C)		

Table 3.3-15 Poly(methyl methacrylate), PMMA: heat capacity, thermal conductivity, and coefficient of thermal expansion

Temperature T (°C)	−200	−150	−100	−50	0	20	50	100	150	200
Heat capacity c_p (kJ/(kg K))		0.67	0.90	1.06	1.26	1.26	1.42	1.85		
Thermal conductivity λ (W/(m K))		0.16	0.18	0.19	0.19	0.19	0.19	0.20	0.19	0.18
Coefficient of thermal expansion α (10^{-5}/K) (linear)		3.0	3.7	4.5	5.7	6.9	7.5	12.0	18.4	

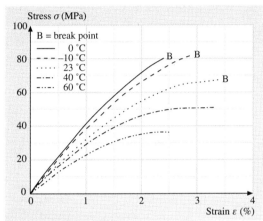

Fig. 3.3-32 Poly(methyl methacrylate), PMMA: stress versus strain

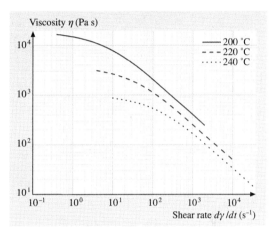

Fig. 3.3-33 Poly(methyl methacrylate), PMMA: viscosity versus shear rate

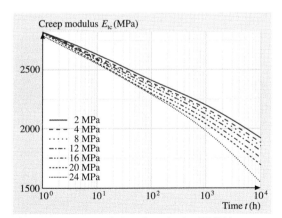

Fig. 3.3-34 Poly(methyl methacrylate), PMMA: creep modulus versus time, at 23 °C

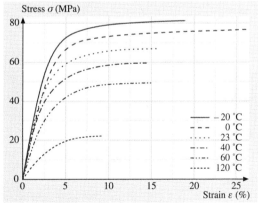

Fig. 3.3-35 Poly(oxymethylene), POM-H: stress versus strain

Table 3.3-16 Poly(oxymethylene), POM-H: heat capacity, thermal conductivity, and coefficient of thermal expansion

Temperature T (°C)	−200	−150	−100	−50	0	20	50	100	150	200
Heat capacity c_p (kJ/(kg K))	0.47	0.65	0.82	1.08	1.27		1.46	1.85		
Thermal conductivity λ (W/(m K))		0.47	0.45	0.43	0.42		0.41			
Coefficient of thermal expansion α (10^{-5}/K) (linear)					9.0	9.5	10.0	16.5	41.0	23.0

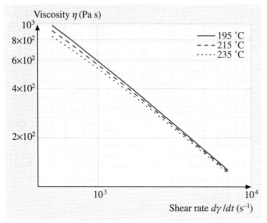

Fig. 3.3-36 Poly(oxymethylene), POM-H: viscosity versus shear rate

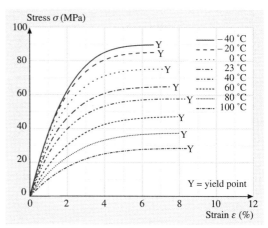

Fig. 3.3-37 Poly(oxymethylene-co-ethylene), POM-R: stress versus strain

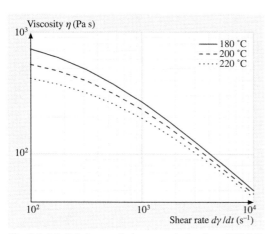

Fig. 3.3-38 Poly(oxymethylene-co-ethylene), POM-R: viscosity versus shear rate

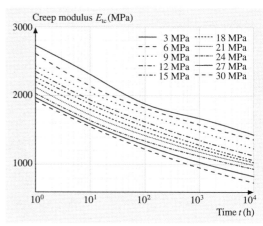

Fig. 3.3-39 Poly(oxymethylene-co-ethylene), POM-R: creep modulus versus time at 23 °C

3.3.3.5 Polyamides

Polyamide 6, PA6; polyamide 66, PA66; polyamide 11, PA11; polyamide 12, PA12; polyamide 610, PA610. Applications: technical parts, bearings, gear wheels, rollers, screws, gaskets, fittings, coverings, housings, automotive parts, houseware, semifinished goods, sports goods, membranes, foils, packings, blow-molded parts, fibers, tanks.

Table 3.3-17 Polyamide 6, PA6; polyamide 66, PA66; polyamide 11, PA11; polyamide 12, PA12; polyamide 610, PA610

	PA6	PA66	PA11	PA12	PA610
Melting temperature T_m (°C)	220–225	255–260	185	175–180	210–220
Enthalpy of fusion ΔH_u (J/mol) (mu)					
Entropy of fusion ΔS_u (J/(K mol)) (mu)					
Heat capacity c_p (kJ/(kg K))	1.7	1.7	1.26	1.26	1.7
Temperature coefficient dc_p/dT (kJ/(kg K^2))					
Enthalpy of combustion ΔH_c (kJ/g)	−31.4	−31.4			
Glass transition temperature T_g (°C)	55	80	50	50	55–60
Vicat softening temperature $T_\mathrm{V}50/50$ (°C)	180–220	195–220	180–190	140–160	205–215
Thermal conductivity λ (W/(m K))	0.29	0.23	0.23	0.23	0.23
Density ϱ (g/cm^3)	1.12–1.14	1.13–1.15	1.04	1.01–1.03	1.06–1.09
Coefficient of thermal expansion α (10^{-5}/K) (linear)	7–10	7–10	13	10–12	8–10
Compressibility κ (1/MPa) (cubic)					
Elastic modulus E (GPa)	2.6–3.2	2.7–3.3	1.0	1.3–1.6	2.0–2.4
Shear modulus G (GPa)	1.1–1.5	1.3–1.7	0.4–0.5	0.5	0.8
Poisson's ratio μ					
Stress at yield σ_y (MPa)	70–90	75–100		45–60	60–70
Strain at yield ε_y (%)	4–5	4.5–5		4–5	4
Stress at 50% elongation σ_{50} (MPa)					
Stress at fracture σ_b (MPa)	70–85	77–84	56	56–65	40
Strain at fracture ε_b (%)	200–300	150–300	500	300	500
Impact strength (Charpy) (kJ/m^2)					
Notched impact strength (Charpy) (kJ/m^2)	3–6	2–3			4–10
Sound velocity v_s (m/s) (longitudinal)	2700	2710			
Sound velocity v_s (m/s) (transverse)	1120	1120			
Shore hardness D	72	75			
Volume resistivity ρ_e (Ω m)	$>10^{13}$	$>10^{12}$	10^{11}	$>10^{13}$	$>10^{13}$
Surface resistivity σ_e (Ω)	$>10^{12}$	$>10^{10}$	10^{11}	$>10^{13}$	$>10^{12}$
Electric strength E_B (kV/mm)	30	25–35	42.5	27–29	
Relative permittivity ε_r (100 Hz)	3.5–4.2	3.2–4	3.7	3.7–4	3.5
Dielectric loss $\tan\delta$ $\tan\delta$ (10^{-4}) (100 Hz)	60–150	50–150	600	300–700	70–150
Refractive index n_D (589 nm)	1.52–1.53	1.52–1.53	1.52–1.53	1.52–1.53	1.53
Temperature coefficient dn_D/dT (1/K)					
Steam permeation (g/(m^2 d)) (100 μm)	10–20	10–20	2.4–4	2.4–4	
Gas permeation (cm^3/(m^2 d bar)) (100 μm)	1–2 (N$_2$)	1–2 (N$_2$)	0.5–0.7 (N$_2$)	0.5–0.7 (N$_2$)	
	2–8 (O$_2$)	2–8 (O$_2$)	2–3.5 (O$_2$)	2–3.5 (O$_2$)	
	80–120 (CO$_2$)	80–120 (CO$_2$)	6–13 (CO)	6–13 (CO)	
Melt viscosity–molar-mass relation					
Viscosity–molar-mass relation					

Table 3.3-18 Polyamide 6, PA6; polyamide 66, PA66; polyamide 610, PA610: heat capacity, thermal conductivity, and coefficient of thermal expansion

Temperature T (°C)	−200	−150	−100	−50	0	20	50	100	150	200
Heat capacity c_p (kJ/(kg K)), PA	0.47	0.73	0.93	1.15	1.38		1.68	2.15	2.60	
Thermal conductivity λ (W/(m K)), PA6		0.29	0.31	0.32	0.32		0.29	0.27	0.25	
Thermal conductivity λ (W/(m K)), PA66		0.32	0.33	0.33	0.33		0.33			
Thermal conductivity λ (W/(m K)), PA610		0.31	0.32	0.33	0.33		0.32	0.31		
Coefficient of thermal expansion α (10^{-5}/K) (linear), PA6			5.0	6.6	8.0	9.1	40.1	15.1	14.0	34.6

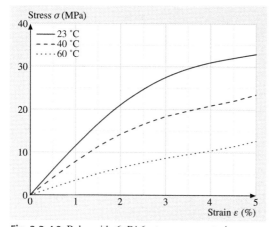

Fig. 3.3-40 Polyamide 6, PA6: stress versus strain

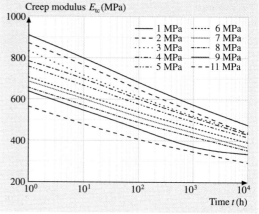

Fig. 3.3-41 Polyamide 6, PA6: creep modulus versus time

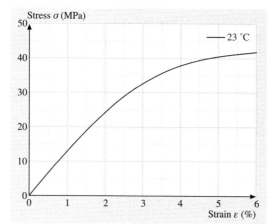

Fig. 3.3-42 Polyamide 66, PA66: stress versus strain

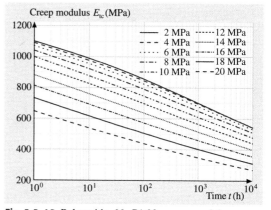

Fig. 3.3-43 Polyamide 66, PA66: creep modulus versus time

3.3.3.6 Polyesters

Polycarbonate, PC. Applications: injection-molded parts, extruded parts, disks, lamp housings, electronic articles, optical articles, houseware, foils.

Poly(ethylene terephthalate), PET. Applications: bearings, gear wheels, shafts, couplings, foils, ribbons, flexible tubing, fibers.

Poly(butylene terephthalate), PBT. Applications: bearings, valve parts, screws, plugs, housings, wheels, houseware.

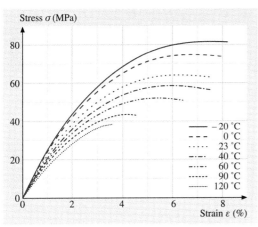

Fig. 3.3-44 Polycarbonate, PC: stress versus strain

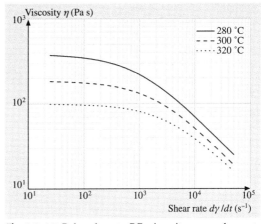

Fig. 3.3-45 Polycarbonate, PC: viscosity versus shear rate

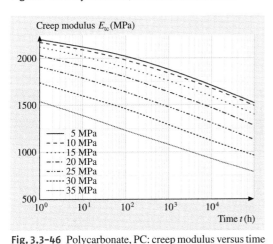

Fig. 3.3-46 Polycarbonate, PC: creep modulus versus time

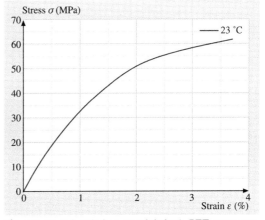

Fig. 3.3-47 Poly(ethylene terephthalate), PET: stress versus strain

Table 3.3-19 Polycarbonate, PC: heat capacity and thermal conductivity

Temperature T (°C)	−200	−150	−100	−50	0	20	50	100	150
Heat capacity c_p (kJ/(kg K))	0.34	0.50	0.70	0.90	1.10	1.17	1.30	1.50	1.70
Thermal conductivity λ (W/(m K))		0.17	0.19	0.21	0.23		0.24	0.24	0.24

Table 3.3-20 Polycarbonate, PC; poly(ethylene terephthalate), PET; poly(butylene terephthalate), PBT; poly(phenylene ether), PPE

	PC	PET	PBT	PPE
Melting temperature T_m (°C)	220–260	250–260 (pcr)	220–225	
Enthalpy of fusion ΔH_u (kJ/mol) (mu)	8.682 (iso), 8.577 (syn)	2.69	54 kJ/kg	
Entropy of fusion ΔS_u (J/(K mol)) (mu)	16.8 (iso), 15.3 (syn)	48.6		
Heat capacity c_p (kJ/(kg K))	1.17	1.223	1.35	
Temperature coefficient dc_p/dT (kJ/(kg K^2))	4.04×10^{-3}			
Enthalpy of combustion ΔH_c (kJ/g)	−30.7	−21.6		
Glass transition temperature T_g (°C)	150	98	60	110–150
Vicat softening temperature $T_V 50/50$ (°C)	83	79	165–180	105–132
Thermal conductivity λ (W/(m K))	0.21	0.11	0.21	0.17–0.22
Density ϱ (g/cm^3)	1.2; 1.38–1.40 (pcr)	1.33–1.35 (am)	1.30–1.32	1.04–1.10
Coefficient of thermal expansion α (10^{-5}/K) (linear)	6.5–7	8 (am), 7 (pcr)	8–10	8–9
Compressibility κ (1/MPa) (cubic)	2.2×10^{-4}	6.99×10^{-6} (melt)		
Elastic modulus E (GPa)	2.3–2.4	2.1–2.4 (am); 2.8–3.1 (pcr)	2.5–2.8	2.0–3.1
Shear modulus G (GPa)	0.72	1.2	1.0	
Poisson's ratio μ		0.38		
Stress at yield σ_y (MPa)	55–65	55 (am), 60–80 (pcr)	50–60	45–70
Strain at yield ε_y (%)	6–7	4 (am), 5–7 (pcr)	3.5–7	3–6
Stress at 50% elongation σ_{50} (MPa)				
Stress at fracture σ_b (MPa)	69–72	30–55	52	35–55
Strain at fracture ε_b (%)	120–125	1.5–3		15–40
Impact strength (Charpy) (kJ/m^2)		13–25		
Notched impact strength (Charpy) (kJ/m^2)		3–5		6–16
Sound velocity v_s (m/s) (longitudinal)	2220	2400		2220
Sound velocity v_s (m/s) (transverse)	909	1150		1000
Shore hardness D			80	
Volume resistivity ρ_e (Ω m)	$> 10^{14}$	$> 10^{13}$	$> 10^{13}$	10^{14}–10^{15}
Surface resistivity σ_e (Ω)	$> 10^{14}$	$> 10^{14}$	$> 10^{14}$	10^{16}–10^{17}
Electric strength E_B (kV/mm)	30–75	42	42	25–35
Relative permittivity ε_r (100 Hz)	2.8–3.2	3.4–3.6	3.3–4.0	2.6–2.9
Dielectric loss $\tan \delta$ (10^{-4}) (100 Hz)	7–20	20	15–20	10–40
Refractive index n_D (589 nm)	1.58–1.59	1.57–1.64	1.58	
Temperature coefficient dn_D/dT (1/K)	-1.42×10^{-4}			
Steam permeation (g/(m^2 d))	4	4.5–5.5, 0.6 (str)		
Gas permeation (cm^3/(m^2 d bar))	680 (N$_2$), 4000 (O$_2$); 14 500 (CO$_2$)	6.6 (N$_2$), 30 (O$_2$); 140 (CO$_2$), 12 (air); 9–15 (str, N$_2$); 80–110 (str, O$_2$); 200–340 (str, CO$_2$)		
Melt viscosity–molar-mass relation				
Viscosity–molar-mass relation		$[\eta] = 4.25 \times 10^{-3} M^{0.69}$ (PET)		

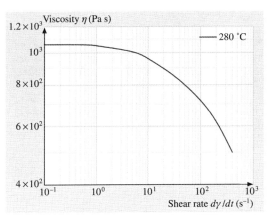

Fig. 3.3-48 Poly(ethylene terephthalate), PET: viscosity versus shear rate

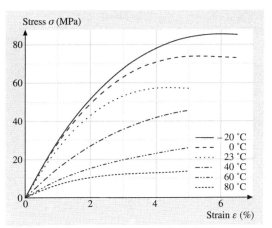

Fig. 3.3-49 Poly(butylene terephthalate), PBT: stress versus strain

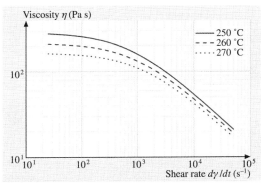

Fig. 3.3-50 Poly(butylene terephthalate), PBT: viscosity versus shear rate

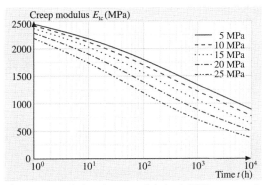

Fig. 3.3-51 Poly(butylene terephthalate), PBT: creep modulus versus time

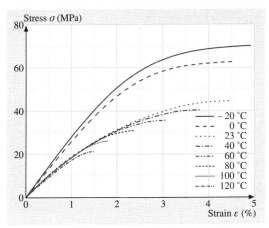

Fig. 3.3-52 Poly(phenylene ether), PPE: stress versus strain

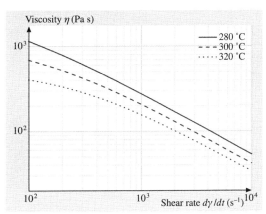

Fig. 3.3-53 Poly(phenylene ether), PPE: viscosity versus shear rate

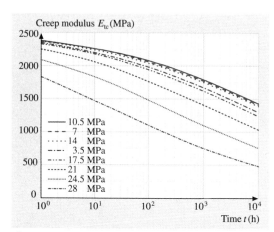

Fig. 3.3-54 Poly(phenylene ether), PPE: creep modulus versus time

3.3.3.7 Polysulfones and Polysulfides

Polysulfone, PSU; Poly(ether sulfone), PES. Applications: injection-molded parts, coatings, electronic articles, printed circuits, houseware, medical devices, membranes, lenses, optical devices.

Poly(phenylene sulfide), PPS. Applications: injection-molded parts, sintered parts, foils, fibers, housings, sockets, foams.

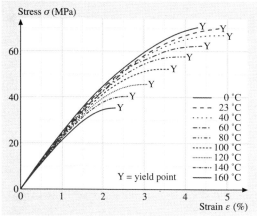

Fig. 3.3-55 Polysulfone, PSU: stress versus strain

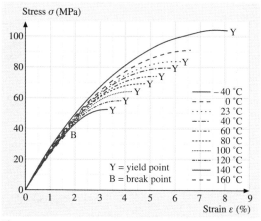

Fig. 3.3-57 Poly(ether sulfone), PES: stress versus strain

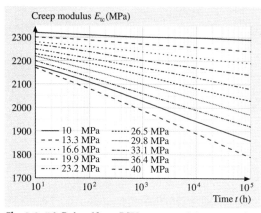

Fig. 3.3-56 Polysulfone, PSU: creep modulus versus time

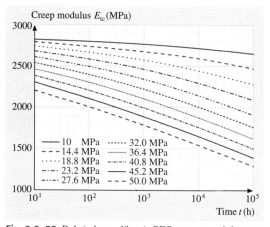

Fig. 3.3-58 Poly(ether sulfone), PES: creep modulus versus time

Table 3.3-21 Polysulfone, PSU; poly(phenylene sulfide), PPS; poly(ether sulfone), PES

	PSU	PPS	PES
Melting temperature T_m (°C)		285	
Enthalpy of fusion ΔH_u (J/mol) (mu)			
Entropy of fusion ΔS_u (J/(K mol)) (mu)			
Heat capacity c_p (kJ/(kg K))	1.30		1.10
Temperature coefficient dc_p/dT (kJ/(kg K^2))			
Enthalpy of combustion ΔH_c (kJ/g)	17		36
Glass transition temperature T_g (°C)	190	85	225
Vicat softening temperature $T_V 50/50$ (°C)	175–210	200	215–225
Thermal conductivity λ (W/(m K))	0.28	0.25	0.18
Density ϱ (g/cm^3)	1.24–1.25	1.34	1.36–1.37
Coefficient of thermal expansion α (10^{-5}/K) (linear)	5.5–6	5.5	5–5.5
Compressibility κ (1/MPa) (cubic)			
Elastic modulus E (GPa)	2.5–2.7	3.4	2.6–2.8
Shear modulus G (GPa)			
Poisson's ratio μ			
Stress at yield σ_y (MPa)	70–80		80–90
Strain at yield ε_y (%)	5.5–6		5.5–6.5
Stress at 50% elongation σ_{50} (MPa)			
Stress at fracture σ_b (MPa)	50–100	75	85
Strain at fracture ε_b (%)	25–30	3	30–80
Impact strength (Charpy) (kJ/m^2)			
Notched impact strength (Charpy) (kJ/m^2)			
Sound velocity v_s (m/s) (longitudinal)	2260		2260
Sound velocity v_s (m/s) (transverse)	920		
Shore hardness D			
Volume resistivity ρ_e (Ω m)	> 10^{13}	> 10^{13}	10^{13}
Surface resistivity σ_e (Ω)	> 10^{15}		> 10^{13}
Electric strength E_B (kV/mm)	20–30	59.5	20–30
Relative permittivity ε_r (100 Hz)	3.2	3.1	3.5–3.7
Dielectric loss tan δ (10^{-4}) (100 Hz)	8–10	4	10–20
Refractive index n_D (589 nm)	1.63		1.65
Temperature coefficient dn_D/dT (1/K)			
Steam permeation (g/(m^2 d)) (25 μm)	6		
Gas permeation (cm^3/(m^2 d bar)) (25 μm)	630 (N$_2$) 3600 (O$_2$) 15 000 (CO$_2$)		
Melt viscosity–molar-mass relation			
Viscosity–molar-mass relation			

3.3.3.8 Polyimides and Polyether Ketones

Poly(amide imide), PAI. Applications: constructional parts, wheels, rotors, bearings, housings, slip fittings, lacquers.

Poly(ether imide), PEI. Applications: electrical housings, sockets, microwave parts, bearings, gear wheels, automotive parts.

Polyimide, PI. Applications: foils, semifinished goods, sintered parts.

Table 3.3-22 Poly(amide imide), PAI; poly(ether imide), PEI; polyimide, PI; poly(ether ether ketone), PEEK

	PAI	PEI	PI	PEEK
Melting temperature T_m (°C)			335–345	
Enthalpy of fusion ΔH_u (J/mol) (mu)				
Entropy of fusion ΔS_u (J/(K mol)) (mu)				
Heat capacity c_p (kJ/(kg K))		1.1		
Temperature coefficient dc_p/dT (kJ/(kg K^2))				
Enthalpy of combustion ΔH_c (kJ/g)				
Glass transition temperature T_g (°C)	240–275	215	250–270	145
Vicat softening temperature T_V50/50 (°C)		220	260	
Thermal conductivity λ (W/(m K))	0.26	0.22	0.29–0.35	0.25
Density ϱ (g/cm^3)	1.38–1.40	1.27	1.43	1.32
Coefficient of thermal expansion α (10^{-5}/K) (linear)	3.0–3.5	5.5–6.0	5–6	4.7
Compressibility κ (1/MPa) (cubic)				
Elastic modulus E (GPa)	4.5–4.7	2.9–3.0	3.0–3.2	4.7
Shear modulus G (GPa)				
Poisson's ratio μ			0.41	
Stress at yield σ_y (MPa)				
Strain at yield ε_y (%)				5
Stress at 50% elongation σ_{50} (MPa)				
Stress at fracture σ_b (MPa)	150–160	105	75–100	100
Strain at fracture ε_b (%)	7–8	60	88	50
Impact strength (Charpy) (kJ/m^2)				
Notched impact strength (Charpy) (kJ/m^2)				
Sound velocity v_s (m/s) (longitudinal)				
Sound velocity v_s (m/s) (transverse)				
Shore hardness D				
Volume resistivity ρ_e (Ω m)	10^{15}	10^{16}	> 10^{14}	10^{14}
Surface resistivity σ_e (Ω)	> 10^{16}	> 10^{16}	> 10^{15}	
Electric strength E_B (kV/mm)	25	25	200	
Relative permittivity ε_r (100 Hz)	3.5–4.2	3.2–3.5	3.4	3.2
Dielectric loss tan δ (10^{-4}) (100 Hz)	10	10–15	52	30
Refractive index n_D (589 nm)		1.66		
Temperature coefficient dn_D/dT (1/K)				
Steam permeation (g/(m^2 d)) (25 µm)			25	
Gas permeation (cm^3/(m^2 d bar)) (25 µm)			94 (N$_2$) 390 (O$_2$) 700 (CO$_2$)	
Melt viscosity–molar-mass relation				
Viscosity–molar-mass relation				

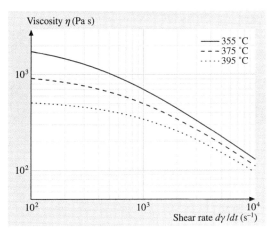

Fig. 3.3-59 Poly(ether imide), PEI: viscosity versus shear rate

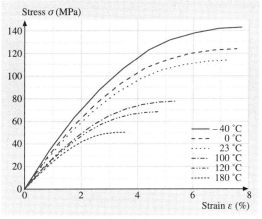

Fig. 3.3-60 Poly(ether imide), PEI: stress versus strain

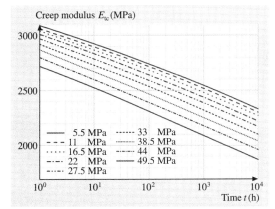

Fig. 3.3-61 Poly(ether imide), PEI: creep modulus versus time

Poly(ether ether ketone), PEEK. Applications: injection-molded parts, automotive parts, aircraft parts, electronic parts, cable insulation, foils, fibers, tapes, plates.

3.3.3.9 Cellulose

Cellulose acetate, CA. Applications: tool handles, pens, combs, buckles, buttons, electronic parts, spectacles, hollow fibers, fibers, foils, lacquers, resin adhesives, sheet molding compounds (SMCs), bulk molding compounds (BMCs).

Cellulose propionate, CP. Applications: fibers, foils, lacquers, resin adhesives, sheet molding compounds, bulk molding compounds (BMC), spectacles, houseware goods, boxes.

Table 3.3-23 Cellulose acetate, CA; cellulose propionate, CP; cellulose acetobutyrate, CAB; ethylcellulose, EC; vulcanized fiber, VF

	CA	CP	CAB	EC	VF
Melting temperature T_m (°C)					
Enthalpy of fusion ΔH_u (J/mol) (mu)					
Entropy of fusion ΔS_u (J/(K mol)) (mu)					
Heat capacity c_p (kJ/(kg K))	1.6	1.7	1.6		
Temperature coefficient dc_p/dT (kJ/(kg K^2))					
Enthalpy of combustion ΔH_c (kJ/g)					
Glass transition temperature T_g (°C)					
Vicat softening temperature T_V 50/50 (°C)	77–110	70–100	65–100		

Table 3.3-23 Cellulose acetate, CA; cellulose propionate, CP; cellulose acetobutyrate, CAB; ... , cont.

	CA	CP	CAB	EC	VF
Thermal conductivity λ (W/(m K))	0.22	0.21	0.21		
Density ϱ (g/cm^3)	1.26–1.32	1.17–1.24	1.16–1.22	1.12–1.15	1.1–1.45
Coefficient of thermal expansion α (10^{-5}/K) (linear)	10–12	11–15	10–15	10	
Compressibility κ (1/MPa) (cubic)					
Elastic modulus E (GPa)	1.0–3.0	1.0–2.4	0.8–2.3	1.2–1.3	
Shear modulus G (GPa)		0.75	0.85		
Poisson's ratio μ					
Stress at yield σ_y (MPa)	25–55	20–50	20–55	35–40	
Strain at yield ε_y (%)	2.5–4	3.5–4.5	3.5–5		
Stress at 50% elongation σ_{50} (MPa)					
Stress at fracture σ_b (MPa)	38	14–55	26		85–100
Strain at fracture ε_b (%)	3	30–100	4		
Impact strength (Charpy) (kJ/m^2)					
Notched impact strength (Charpy) (kJ/m^2)					
Sound velocity v_s (m/s) (longitudinal)					
Sound velocity v_s (m/s) (transverse)					
Shore hardness D					
Volume resistivity ρ_e (Ω m)	10^{10}–10^{14}	10^{10}–10^{14}	10^{10}–10^{14}	10^{11}–10^{13}	
Surface resistivity σ_e (Ω)	10^{10}–10^{14}	10^{12}–10^{14}	10^{12}–10^{14}	10^{11}–10^{13}	
Electric strength E_B (kV/mm)	25–35	30–35	32–35	≈ 30	
Relative permittivity ε_r (100 Hz)	5–6	4.0–4.2	3.7–4.2	≈ 4	
Dielectric loss tan δ (10^{-4}) (100 Hz)	70–100	50	50–70	100	
Refractive index n_D (589 nm)	1.47–1.50	1.47–1.48	1.48		
Temperature coefficient dn_D/dT (1/K)					
Steam permeation (g/(m^2 d)) (25 μm)	150–600		460–600		
Gas permeation (cm^3/(m^2 d bar)) (25 μm)	470–630 (N$_2$) 13 000–15 000 (O$_2$) 14 000 (CO$_2$) 1800–2300 (air)		3800 (N$_2$) 15 000 (O$_2$) 94 000 (CO$_2$)		
Melt viscosity–molar-mass relation					
Viscosity–molar-mass relation					

Cellulose acetobutyrate, CAB. Applications: fibers, foils, lacquers, resin adhesives, sheet molding compounds, bulk molding compounds, automotive parts, switches, light housings, spectacles.

Ethylcellulose, EC. Applications: foils, injection-molded parts, lacquers, adhesives.

Vulcanized fiber, VF. Applications: gear wheels, abrasive wheels, case plates.

3.3.3.10 Polyurethanes

Polyurethane, PUR; thermoplastic polyurethane elastomer, TPU. Applications: blocks, plates, formed pieces, cavity forming, composites, coatings, impregnations, films, foils, fibers, molded foam plastics, insulation boards, sandwich boards, installation boards, reaction injection molding (RIM), automotive parts, houseware goods, sports goods.

Table 3.3-24 Polyurethane, PUR; thermoplastic polyurethane elastomer, TPU

	PUR	TPU
Melting temperature T_m (°C)		
Enthalpy of fusion ΔH_u (J/mol) (mu)		
Entropy of fusion ΔS_u (J/(K mol)) (mu)		
Heat capacity c_p (kJ/(kg K))	1.76	0.5
Temperature coefficient dc_p/dT (kJ/(kg K^2))		
Enthalpy of combustion ΔH_c (kJ/mol) (mu)		
Glass transition temperature T_g (K)	15–90	−40
Vicat softening temperature T_V 50/50 (°C)	100–180	
Thermal conductivity λ (W/(m K))	0.58	1.70
Density ϱ (g/cm^3)	1.05	1.1–1.25
Coefficient of thermal expansion α (10^{-5}/K) (linear)	1.0–2.0	15
Compressibility κ (1/MPa) (cubic)		
Elastic modulus E (GPa)	4.0	0.015–0.7
Shear modulus G (GPa)		0.006–0.23
Poisson's ratio μ		
Stress at yield σ_y (MPa)		
Strain at yield ε_y (%)		
Stress at 50% elongation σ_{50} (MPa)		
Stress at fracture σ_b (MPa)	70–80	30–50
Strain at fracture ε_b (%)	3–6	300–500
Impact strength (Charpy) (kJ/m^2)		
Notched impact strength (Charpy) (kJ/m^2)		
Sound velocity v_s (m/s) (longitudinal)	1550–1750	
Sound velocity v_s (m/s) (transverse)		
Shore hardness D	60–90	30–70
Volume resistivity ρ_e (Ω m)	10^{14}	10^{10}
Surface resistivity σ_e (Ω)	10^{14}	10^{11}
Electric strength E_B (kV/mm)	24	30–60
Relative permittivity ε_r (100 Hz)	3.6	6.5
Dielectric loss tan δ (10^{-4}) (100 Hz)	500	300
Refractive index n_D (589 nm)		
Temperature coefficient dn_D/dT (1/K)		
Steam permeation (g/(m^2 d)) (25 μm)		13–25
Gas permeation (cm^3/(m^2 d bar)) (25 μm)		550–1600 (N$_2$) 1000–4500 (O$_2$) 6000–22 000 (CO$_2$)
Melt viscosity–molar-mass relation		
Viscosity–molar-mass relation		

Table 3.3-25 Polyurethane, PUR: thermal conductivity and coefficient of thermal expansion

Temperature T (°C)	−200	−150	−100	−50	0	20	50	100
Thermal conductivity λ (W/(m K))		0.20	0.21	0.22	0.21		0.20	0.20
Average coefficient of thermal expansion α^* (10^{-5}/K) (linear)[a]	9.9	12.7	16.0	26.0	20.0			

[a] $\alpha^*(T) = (1/(T-20)) \int_{T=20}^{T'} \alpha(T)\, dT$, T in °C

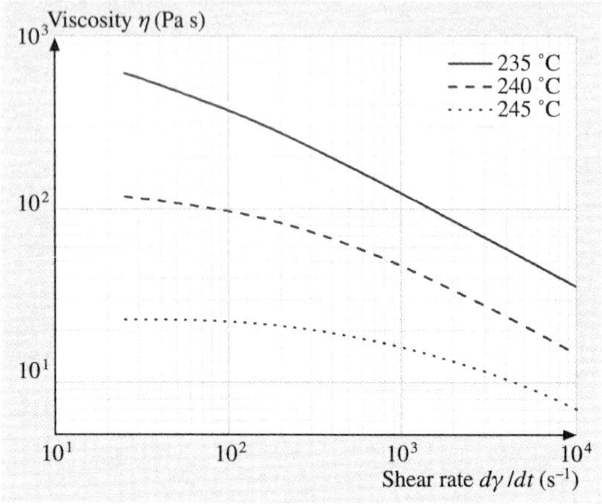

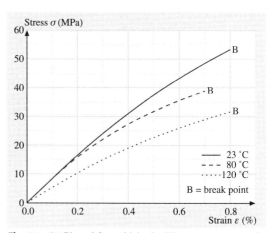

Fig. 3.3-62 Thermoplastic polyurethane elastomer, TPU: viscosity versus shear rate

Fig. 3.3-63 Phenol formaldehyde, PF: stress versus strain

3.3.3.11 Thermosets

Phenol formaldehyde, PF; urea formaldehyde, UF; melamine formaldehyde, MF. Applications: molded pieces, granulated molding compounds (GMCs), bulk molding compounds, dough molding compounds (DMCs), sheet molding compounds, plates, gear wheels, abrasive wheels, lacquers, coatings, chipboards, adhesives, foams, fibers.

Unsaturated polyester, UP; diallyl phthalate, DAP. Applications: molding compounds, cast resins, electronic parts, automotive parts, containers, tools, houseware goods, sports goods.

Table 3.3-26 Phenol formaldehyde, PF; urea formaldehyde, UF; melamine formaldehyde, MF

	PF	UF	MF
Melting temperature T_m (°C)			
Enthalpy of fusion ΔH_u (J/mol) (mu)			
Entropy of fusion ΔS_u (J/(K mol)) (mu)			
Heat capacity c_p (kJ/(kg K))	1.30	1.20	1.20
Temperature coefficient dc_p/dT (kJ/(kg K^2))			
Enthalpy of combustion ΔH_c (kJ/g)			
Glass transition temperature T_g (°C)			
Vicat softening temperature $T_V 50/50$ (°C)			
Thermal conductivity λ (W/(m K))	0.35	0.40	0.50
Density ϱ (g/cm^3)	1.4	1.5	1.5
Coefficient of thermal expansion α (10^{-5}/K) (linear)	3–5	5–6	5–6

Table 3.3-26 Phenol formaldehyde, PF; urea formaldehyde, UF; melamine formaldehyde, MF, cont.

	PF	UF	MF
Compressibility κ (1/MPa) (cubic)			
Elastic modulus E (GPa)	5.6–12	7.0–10.5	4.9–9.1
Shear modulus G (GPa)			
Poisson's ratio μ			
Stress at yield σ_y (MPa)			
Strain at yield ε_y (%)			
Stress at 50% elongation σ_{50} (MPa)			
Stress at fracture σ_b (MPa)	25	30	30
Strain at fracture ε_b (%)	0.4–0.8	0.5–1.0	0.6–0.9
Impact strength (Charpy) (kJ/m^2)	4.5–10		
Notched impact strength (Charpy) (kJ/m^2)	1.5–3		
Sound velocity v_s (m/s) (longitudinal)			
Sound velocity v_s (m/s) (transverse)			
Shore hardness D	82		90
Volume resistivity ρ_e (Ω m)	10^9	10^9	10^9
Surface resistivity σ_e (Ω)	$> 10^8$	$> 10^{10}$	$> 10^8$
Electric strength E_B (kV/mm)	30–40	30–40	29–30
Relative permittivity ε_r (100 Hz)	6	8	9
Dielectric loss tan δ (10^{-4}) (100 Hz)	1000	400	600
Refractive index n_D (589 nm)	1.63		
Temperature coefficient dn_D/dT (1/K)			
Steam permeation (g/(m^2 d)) (40 μm)	43		400
Gas permeation (cm^3/(m^2 d bar))			
Melt viscosity–molar-mass relation			
Viscosity–molar-mass relation			

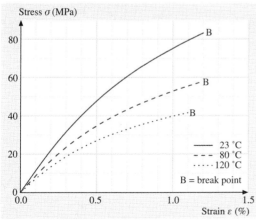

Fig. 3.3-64 Melamine formaldehyde, MF: stress versus strain

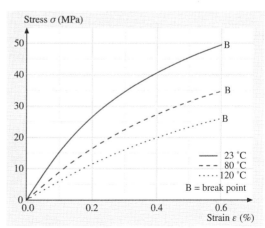

Fig. 3.3-65 Unsaturated polyester, UP: stress versus strain

Table 3.3-27 Unsaturated polyester, UP; diallyl phthalate, DAP; silicone resin, SI; epoxy resin, EP

	UP	DAP	SI	EP
Melting temperature T_m (°C)				
Enthalpy of fusion ΔH_u (J/mol) (mu)				
Entropy of fusion ΔS_u (J/(K mol)) (mu)				
Heat capacity c_p (kJ/(kg K))	1.20		0.8–0.9	0.8–1.0
Temperature coefficient dc_p/dT (kJ/(kg K^2))				
Enthalpy of combustion ΔH_c (kJ/g)				
Glass transition temperature T_g (°C)	70–120			
Vicat softening temperature T_V 50/50 (°C)				
Thermal conductivity λ (W/(m K))	0.7	0.6	0.3–0.4	0.88
Density ϱ (g/cm^3)	2.0	1.51–1.78	1.8–1.9	1.9
Coefficient of thermal expansion α (10^{-5}/K) (linear)	2–4	1–3.5	2–5	1.1–3.5
Compressibility κ (1/MPa) (cubic)				
Elastic modulus E (GPa)	14–20	9.8–15.5	6–12	21.5
Shear modulus G (GPa)				
Poisson's ratio μ				
Stress at yield σ_y (MPa)				
Strain at yield ε_y (%)				
Stress at 50% elongation σ_{50} (MPa)				
Stress at fracture σ_b (MPa)	30	40–75	28–46	30–40
Strain at fracture ε_b (%)	0.6–1.2			4
Impact strength (Charpy) (kJ/m^2)				
Notched impact strength (Charpy) (kJ/m^2)				
Sound velocity v_s (m/s) (longitudinal)				
Sound velocity v_s (m/s) (transverse)				
Shore hardness D	82			
Volume resistivity ρ_e (Ω m)	$> 10^{10}$	10^{11}–10^{14}	10^{12}	$> 10^{12}$
Surface resistivity σ_e (Ω)	$> 10^{10}$	10^{13}	10^{12}	$> 10^{12}$
Electric strength E_B (kV/mm)	25–53	40	20–40	30–40
Relative permittivity ε_r (100 Hz)	6	5.2	4	3.5–5
Dielectric loss tan δ (10^{-4}) (100 Hz)	400	400	300	10
Refractive index n_D (589 nm)	1.54–1.58			1.47
Temperature coefficient dn_D/dT (1/K)				
Steam permeation (g/(m^2 d))				
Gas permeation (cm^3/(m^2 d bar))				
Melt viscosity–molar-mass relation				
Viscosity–molar-mass relation				

Table 3.3-28 Unsaturated polyester, UP: coefficient of thermal expansion

Temperature T (°C)	−200	−150	−100	−50	0	20	50	100
Coefficient of thermal expansion α (10^{-5}/K) (linear)	3.0	4.1	4.9	5.8	7.3	8.4	10.7	15.0

Table 3.3-29 Epoxy resin, EP: thermal conductivity and coefficient of thermal expansion

Temperature T (°C)	−200	−150	−100	−50	0	20	50	100	150
Thermal conductivity λ (W/(m K))					0.20	0.20	0.20	0.20	
Coefficient of thermal expansion α (10^{-5}/K) (linear)	1.8	2.8	3.8	4.9	6.1	6.2	6.3	7.5	13.0

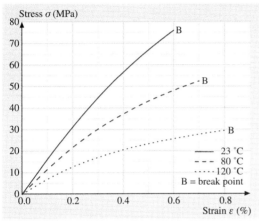

Fig. 3.3-66 Epoxy resin, EP: stress versus strain

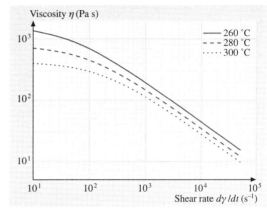

Fig. 3.3-67 Poly(acrylonitrile-co-butadiene-co-styrene) + polycarbonate, ABS + PC: viscosity versus shear rate

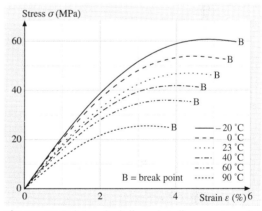

Fig. 3.3-68 Poly(acrylonitrile-co-butadiene-co-styrene) + polycarbonate, ABS + PC: stress versus strain

Silicone resin, SI. Applications: molding compounds, laminates.

Epoxy resin EP, Applications: moulding compounds.

3.3.3.12 Polymer Blends

Polypropylene + ethylene/propylene/diene rubber, PP + EPDM. Applications: automotive parts, flexible tubes, sports goods, toys.

Poly(acrylonitrile-co-butadiene-co-styrene) + polycarbonate, ABS + PC; Poly(acrylonitrile-co-butadiene-co-acrylester) + polycarbonate ASA + PC; poly(acrylonitrile-co-butadiene-co-styrene) + polyamide, ABS + PA. Applications: semifinished goods, automotive parts, electronic parts, optical parts, houseware goods.

Poly(vinyl chloride) + poly(vinyl chloride-co-acrylate), PVC + VC/A; poly(vinyl chloride) + chlorinated polyethylene, PVC + PE-C, poly(vinyl chloride) + poly(acrylonitrile-co-butadiene-co-acrylester), PVC + ASA. Applications: semifinished goods, foils, plates, profiles, pipes, fittings, gutters, window frames, door frames, panels, housings, bottles, blow molding, disks, blocking layers, fibers, fleeces, nets.

Polycarbonate + poly(ethylene terephthalate), PC + PET; polycarbonate + liquid crystal polymer, PC + LCP; polycarbonate + poly(butylene terephthalate), PC + PBT. Applications: optical devices, covering devices, panes, safety glasses, semifinished goods, houseware goods, compact discs, bottles.

Poly(ethylene terephthalate) + polystyrene, PET + PS; poly(butylene terephthalate) + polystyrene, PBT + PS. Applications: same as for PET, PBT, and PS.

Table 3.3-30 Polypropylene + ethylene/propylene/diene rubber, PP + EPDM; poly(acrylonitrile-co-butadiene-co-styrene) + polycarbonate, ABS + PC; poly(acrylonitrile-co-butadiene-co-styrene) + polyamide, ABS + PA; poly(acrylonitrile-co-butadiene-co-acrylester) + polycarbonate, ASA + PC

	PP + EPDM	ABS + PC	ABS + PA	ASA + PC
Melting temperature T_m (°C)	160–168			
Enthalpy of fusion ΔH_u (J/mol) (mu)				
Entropy of fusion ΔS_u (J/(K mol)) (mu)				
Heat capacity c_p (kJ/(kg K))				
Temperature coefficient dc_p/dT (kJ/(kg K^2))				
Enthalpy of combustion ΔH_c (kJ/g)				
Glass transition temperature T_g (°C)				
Vicat softening temperature T_V 50/50 (°C)				
Thermal conductivity λ (W/(m K))				
Density ϱ (g/cm^3)	0.89–0.92	1.08–1.17	1.07–1.09	1.15
Coefficient of thermal expansion α (10^{-5}/K) (linear)	15–18	7–8.5	9	7–9
Compressibility κ (1/MPa) (cubic)				
Elastic modulus E (GPa)	0.5–1.2	2.0–2.6	1.2–1.3	2.3–2.6
Shear modulus G (GPa)				
Poisson's ratio μ				
Stress at yield σ_y (MPa)	10–25	40–60	30–32	53–63
Strain at yield ε_y (%)	10–35	3–3.5		4.6–5
Stress at 50% elongation σ_{50} (MPa)				
Stress at fracture σ_b (MPa)				
Strain at fracture ε_b (%)	> 50	> 50	> 50	> 50
Impact strength (Charpy) (kJ/m^2)				
Notched impact strength (Charpy) (kJ/m^2)				
Sound velocity v_s (m/s) (longitudinal)				
Sound velocity v_s (m/s) (transverse)				
Shore hardness D				
Volume resistivity ρ_e (Ω m)	> 10^{14}	> 10^{14}	2 × 10^{12}	10^{11}–10^{13}
Surface resistivity σ_e (Ω)	> 10^{13}	10^{14}	3 × 10^{14}	10^{13}–10^{14}
Electric strength E_B (kV/mm)	35–40	24	30	
Relative permittivity ε_r (100 Hz)	2.3	3		3–3.5
Dielectric loss tan δ (10^{-4}) (100 Hz)	2.5	30–60		20–160
Refractive index n_D (589 nm)				
Temperature coefficient dn_D/dT (1/K)				
Steam permeation (g/(m^2 d))				
Gas permeation (cm^3/(m^2 d bar))				
Melt viscosity–molar-mass relation				
Viscosity–molar-mass relation				

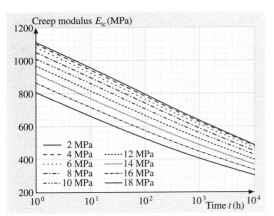

Fig. 3.3-69 Poly(acrylonitrile-co-butadiene-co-styrene) + polycarbonate, ABS + PC: creep modulus versus time

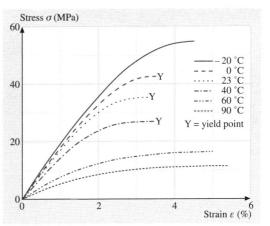

Fig. 3.3-70 Poly(acrylonitrile-co-butadiene-co-styrene) + polyamide, ABS + PA: stress versus strain

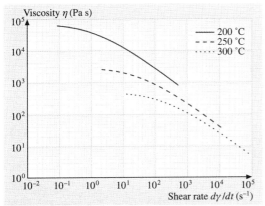

Fig. 3.3-71 Poly(acrylonitrile-co-butadiene-co-acrylester) + polycarbonate, ASA + PC: viscosity versus shear rate

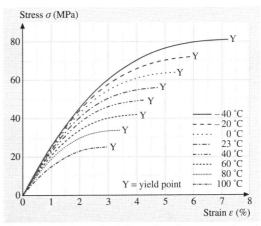

Fig. 3.3-72 Poly(acrylonitrile-co-butadiene-co-acrylester) + polycarbonate, ASA + PC: stress versus strain

Poly(butylene terephthalate) + poly(acrylonitrile-co-butadiene-co-acrylester), PBT + ASA. Applications: same as for PBT and ASA.

Polysulfone + poly(acrylonitrile-co-butadiene-co-styrene), PSU + ABS. Applications: same as for PSU.

Poly(styrene-co-butadiene), PPE + SB. Applications: same as for PPE.

Poly(phenylene ether) + polyamide 66, PPE + PA66. Applications: automotive parts, semifinished goods.

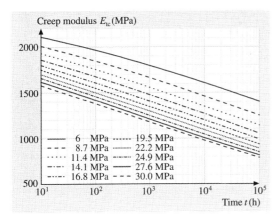

Fig. 3.3-73 Poly(acrylonitrile-co-butadiene-co-acrylester) + polycarbonate, ASA + PC: creep modulus versus time

Table 3.3-31 Poly(vinyl chloride) + poly(vinyl chloride-co-acrylate), PVC + VC/A; poly(vinyl chloride) + chlorinated polyethylene, PVC + PE-C; poly(vinyl chloride) + poly(acrylonitrile-co-butadiene-co-acrylester), PVC + ASA

	PVC + VC/A	PVC + PE-C	PVC + ASA
Melting temperature T_m (°C)			
Enthalpy of fusion ΔH_u (J/mol) (mu)			
Entropy of fusion ΔS_u (J/(K mol)) (mu)			
Heat capacity c_p (kJ/(kg K))			
Temperature coefficient dc_p/dT (kJ/(kg K^2))			
Enthalpy of combustion ΔH_c (kJ/g)			
Glass transition temperature T_g (°C)			
Vicat softening temperature $T_V 50/50$ (°C)			
Thermal conductivity λ (W/(m K))			
Density ϱ (g/cm^3)	1.42–1.44	1.36–1.43	1.28–1.33
Coefficient of thermal expansion α (10^{-5}/K) (linear)	7–7.5	8	7.5–10
Compressibility κ (1/MPa) (cubic)			
Elastic modulus E (GPa)	2.5–2.7	2.6	2.6–2.8
Shear modulus G (GPa)			
Poisson's ratio μ			
Stress at yield σ_y (MPa)	45	40–50	45–55
Strain at yield ε_y (%)	4–5	3	3–3.5
Stress at 50% elongation σ_{50} (MPa)			
Stress at fracture σ_b (MPa)			
Strain at fracture ε_b (%)	> 50	10 to >50	≈ 8
Impact strength (Charpy) (kJ/m^2)			
Notched impact strength (Charpy) (kJ/m^2)			
Sound velocity v_s (m/s) (longitudinal)			
Sound velocity v_s (m/s) (transverse)			
Shore hardness D			
Volume resistivity ρ_e (Ω m)	10^{13}	10^{12} to >10^{13}	10^{12} to >10^{14}
Surface resistivity σ_e (Ω)	> 10^{13}	> 10^{13}	10^{12}–10^{14}
Electric strength E_B (kV/mm)			
Relative permittivity ε_r (100 Hz)	3.5	3.1	3.7–4.3
Dielectric loss tan δ (10^{-4}) (100 Hz)	120	140	100–120
Refractive index n_D (589 nm)			
Temperature coefficient dn_D/dT (1/K)			
Steam permeation (g/(m^2 d))			
Gas permeation (cm^3/(m^2 d bar))			
Melt viscosity–molar-mass relation			
Viscosity–molar-mass relation			

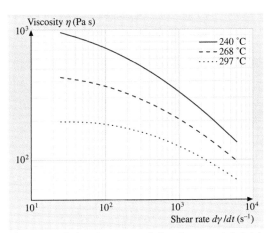

Fig. 3.3-74 Polycarbonate + poly(butylene terephthalate), PC + PBT: viscosity versus shear rate

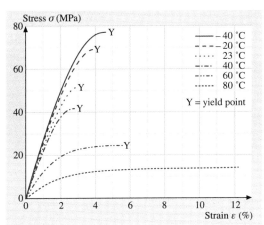

Fig. 3.3-75 Polycarbonate + poly(butylene terephthalate), PC + PBT: stress versus strain

Table 3.3-32 Polycarbonate + poly(ethylene terephthalate), PC + PET; polycarbonate + liquid crystal polymer, PC + LCP; polycarbonate + poly(butylene terephthalate), PC + PBT; poly(ethylene terephthalate) + polystyrene, PET + PS; poly(butylene terephthalate) + polystyrene, PBT + PS

	PC + PET	PC + LCP	PC + PBT	PET + PS	PBT + PS
Melting temperature T_m (°C)					
Enthalpy of fusion ΔH_u (J/mol) (mu)					
Entropy of fusion ΔS_u (J/(K mol)) (mu)					
Heat capacity c_p (kJ/(kg K))				1.05	1.30
Temperature coefficient dc_p/dT (kJ/(kg K^2))					
Enthalpy of combustion ΔH_c (kJ/g)					
Glass transition temperature T_g (°C)					
Vicat softening temperature T_V 50/50 (°C)					
Thermal conductivity λ (W/(m K))				0.24	0.21
Density ϱ (g/cm^3)	1.22		1.2–1.26	1.37	1.31
Coefficient of thermal expansion α (10^{-5}/K) (linear)	9–10		8–9	7	6.0
Compressibility κ (1/MPa) (cubic)					
Elastic modulus E (GPa)	2.1–2.3	2.6–4.0	2.3	3.1	2.0
Shear modulus G (GPa)					
Poisson's ratio μ					
Stress at yield σ_y (MPa)	50–55	66	50–60	47	40
Strain at yield ε_y (%)	5	5.6–6.9	4–5		
Stress at 50% elongation σ_{50} (MPa)					
Stress at fracture σ_b (MPa)		74–82			15
Strain at fracture ε_b (%)	> 50		25 to > 50	50	
Impact strength (Charpy) (kJ/m^2)					
Notched impact strength (Charpy) (kJ/m^2)					
Sound velocity v_s (m/s) (longitudinal)					

Table 3.3-32 Polycarbonate + poly(ethylene terephthalate), PC + PET; polycarbonate + liquid crystal polymer, PC + LCP; polycarbonate + poly(butylene terephthalate), PC + PBT; poly(ethylene terephthalate) + polystyrene, PET + PS; poly(butylene terephthalate) + polystyrene, PBT + PS, cont.

	PC + PET	PC + LCP	PC + PBT	PET + PS	PBT + PS
Sound velocity v_s (m/s) (transverse)					
Shore hardness D					
Volume resistivity ρ_e (Ω m)	$> 10^{13}$		10^{14}	$> 10^{12}$	
Surface resistivity σ_e (Ω)	$> 10^{15}$		10^{14}	$> 10^{12}$	
Electric strength E_B (kV/mm)	30		35		
Relative permittivity ε_r (100 Hz)	3.3		3.3		
Dielectric loss $\tan\delta$ (10^{-4}) (100 Hz)	200		20–40		
Refractive index n_D (589 nm)					
Temperature coefficient dn_D/dT (1/K)					
Steam permeation (g/(m² d))					
Gas permeation (cm³/(m² d bar))					
Melt viscosity–molar-mass relation					
Viscosity–molar-mass relation					

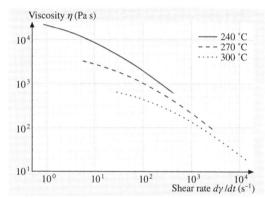

Fig. 3.3-76 Poly(phenylene ether) + poly(styrene-co-butadiene), PPE + SB: viscosity versus shear rate

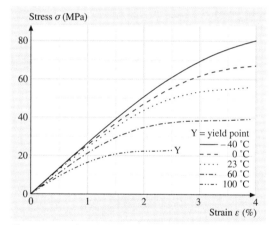

Fig. 3.3-77 Poly(phenylene ether) + poly(styrene-co-butadiene), PPE + SB: stress versus strain

Poly(phenylene ether) + polystyrene, PPE + PS. Applications: automotive parts, foams, office goods, electrical parts, houseware goods.

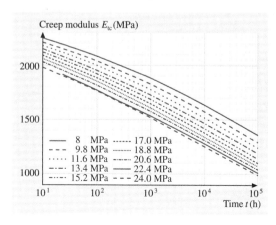

Fig. 3.3-78 Poly(phenylene ether) + poly(styrene-co-butadiene), PPE + SB: creep modulus versus time

Table 3.3-33 Poly(butylene terephthalate) + poly(acrylonitrile-co-butadiene-co-acrylester), PBT + ASA; polysulfon + poly(acrylonitrile-co-butadiene-co-styrene), PSU + ABS; poly(phenylene ether) + poly(styrene-co-butadiene), PPE + SB; poly(phenylene ether) + polyamide 66, PPE + PA66; poly(phenylene ether) + polystyrene PPE + PS

	PBT + ASA	PSU + ABS	PPE + SB	PPE + PA66	PPE + PS
Melting temperature T_m (°C)	225				
Enthalpy of fusion ΔH_u (J/mol) (mu)					
Entropy of fusion ΔS_u (J/(K mol)) (mu)					
Heat capacity c_p (kJ/(kg K))					1.40
Temperature coefficient dc_p/dT (kJ/(kg K^2))					
Enthalpy of combustion ΔH_c (kJ/g)					
Glass transition temperature T_g (°C)					
Vicat softening temperature $T_V 50/50$ (°C)					
Thermal conductivity λ (W/(m K))					0.23
Density ϱ (g/cm^3)	1.21–1.22	1.13	1.04–1.06	1.09–1.10	1.06
Coefficient of thermal expansion α (10^{-5}/K) (linear)	10	6.5	6.0–7.5	8–11	6.0
Compressibility κ (1/MPa) (cubic)					
Elastic modulus E (GPa)	2.5	2.1	1.9–2.7	2.0–2.2	2.5
Shear modulus G (GPa)					
Poisson's ratio μ					
Stress at yield σ_y (MPa)	53	50	45–65	50–60	55
Strain at yield ε_y (%)	3.6	4	3–7	5	
Stress at 50% elongation σ_{50} (MPa)					
Stress at fracture σ_b (MPa)			20 to >50	>50	
Strain at fracture ε_b (%)	>50	>50			50
Impact strength (Charpy) (kJ/m^2)					
Notched impact strength (Charpy) (kJ/m^2)					
Sound velocity v_s (m/s) (longitudinal)					
Sound velocity v_s (m/s) (transverse)					
Shore hardness D					
Volume resistivity ρ_e (Ω m)	>10^{14}	>10^{13}	>10^{14}	>10^{11}	10^{12}
Surface resistivity σ_e (Ω)	>10^{15}	>10^{14}	>10^{14}	>10^{12}	10^{12}
Electric strength E_B (kV/mm)	30	20–30	35–40	95	
Relative permittivity ε_r (100 Hz)	3.3	3.1–3.3	2.6–2.8	3.1–3.4	
Dielectric loss tan δ (10^{-4}) (100 Hz)	200	40–50	5–15	450	
Refractive index n_D (589 nm)					
Temperature coefficient dn_D/dT (1/K)					
Steam permeation (g/(m^2 d))					
Gas permeation (cm^3/(m^2 d bar))					
Melt viscosity–molar-mass relation					
Viscosity–molar-mass relation					

References

3.1 M. D. Lechner, K. Gehrke, E. H. Nordmeier: *Makromolekulare Chemie* (Birkhäuser, Basel 2003)

3.2 G. W. Becker, D. Braun (Eds.): *Kunststoff Handbuch* (Hanser, Munich 1969–1990)

3.3 Kunststoffdatenbank CAMPUS, M-Base, Aachen (www.m-base.de)

3.4 J. Brandrup, E. H. Immergut, E. A. Grulke (Eds.): *Polymer Handbook* (Wiley, New York 1999)

3.5 J. E. Mark (Ed.): *Physical Properties of Polymers Handbook* (AIP, Woodbury 1996)

3.6 C. C. Ku, R. Liepins: *Electrical Properties of Polymers* (Hanser, Munich 1987)

3.7 H. Saechtling: *Kunststoff Taschenbuch* (Hanser, Munich 1998)

3.8 W. Hellerich, G. Harsch, S. Haenle: *Werkstoff-Führer Kunststoffe* (Hanser, Munich 1996)

3.9 B. Carlowitz: *Kunststoff Tabellen* (Hanser, Munich 1995)

3.10 G. Allen, J. C. Bevington (Eds.): *Comprehensive Polymer Science* (Pergamon, Oxford 1989)

3.11 H. Domininghaus: *Die Kunststoffe und ihre Eigenschaften* (VDI, Düsseldorf 1992)

3.12 H. J. Arpe (Ed.): *Ullmann's Encyclopedia of Industrial Chemistry*, 5th edn. (VCH, Weinheim 1985–1996)

3.13 R. E. Kirk, D. F. Othmer (Eds.): *Encyclopedia of Chemical Technology*, 4th edn. (Wiley, New York 1978–1984)

3.14 H. F. Mark, N. Bikales, C. G. Overberger, G. Menges, J. I. Kroschwitz: *Encyclopedia of Polymer Science and Engineering* (Wiley, New York 1985–1990)

3.15 Werkstoff-Datenbank POLYMAT, Deutsches Kunststoff-Institut, Darmstadt

3.16 M. Neubronner: Stoffwerte von Kunststoffen. In: *VDI-Wärmeatlas*, 8th edn. (VDI, Düsseldorf 1997)

3.4. Glasses

This chapter has been conceived as a source of information for scientists, engineers, and technicians who need data and commercial-product information to solve their technical task by using glasses as engineering materials. It is not intended to replace the comprehensive scientific literature. The fundamentals are merely sketched, to provide a feeling for the unique behavior of this widely used class of materials.

The properties of glasses are as versatile as their composition. Therefore only a selection of data can be listed, but their are intended to cover the preferred glass types of practical importance. Wherever possible, formulas, for example for the optical and thermal properties, are given with their correct constants, which should enable the reader to calculate the data needed for a specific situation by him/herself.

For selected applications, the suitable glass types and the main instructions for their processing are presented. Owing to the availability of the information, the products of Schott AG have a certain preponderance here. The properties of glass types from other manufacturers have been included whenever available.

3.4.1 **Properties of Glasses – General Comments** 526
3.4.2 **Composition and Properties of Glasses** . 527
3.4.3 **Flat Glass and Hollowware** 528
  3.4.3.1 Flat Glass 528
  3.4.3.2 Container Glass 529
3.4.4 **Technical Specialty Glasses** 530
  3.4.4.1 Chemical Stability of Glasses 530
  3.4.4.2 Mechanical and Thermal Properties 533
  3.4.4.3 Electrical Properties 537
  3.4.4.4 Optical Properties 539
3.4.5 **Optical Glasses** 543
  3.4.5.1 Optical Properties 543
  3.4.5.2 Chemical Properties 549
  3.4.5.3 Mechanical Properties 550
  3.4.5.4 Thermal Properties 556
3.4.6 **Vitreous Silica** 556
  3.4.6.1 Properties of Synthetic Silica 556
  3.4.6.2 Gas Solubility and Molecular Diffusion 557
3.4.7 **Glass–Ceramics** 558
3.4.8 **Glasses for Miscellaneous Applications** . 559
  3.4.8.1 Sealing Glasses 559
  3.4.8.2 Solder and Passivation Glasses .. 562
  3.4.8.3 Colored Glasses 565
  3.4.8.4 Infrared-Transmitting Glasses ... 568
References 572

Glasses are very special materials that are formed only under suitable thermodynamic conditions; these conditions may be natural or man-made. Most glass products manufactured on a commercial scale are made by quenching a mixture of oxides from the melt (Fig. 3.4-1).

For some particular applications, glasses are also made by other technologies, for example by chemical vapor deposition to achieve extreme purity, as required in optical fibers for communication, or by roller chilling in the case of amorphous metals, which need extremely high quenching rates. The term "amorphous" is a more general, generic expression in comparison with the term "glass". Many different technological routes are described in [4.1].

Glasses are also very universal engineering materials. Variation of the composition results in a huge variety of glass types, families, or groups, and a corresponding variety of properties. In large compositional areas, the properties depend continuously on composition, thus allowing one to design a set of properties to fit a specific application. In narrow ranges, the properties depend linearly on composition; in wide ranges, nonlinearity and step-function behavior have to be con-

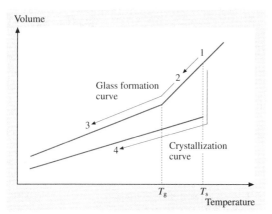

Fig. 3.4-1 Schematic volume–temperature curves for glass formation along path 1–2–3 and crystallization along path 1–4. T_s, melting temperature; T_g, transformation temperature. 1, liquid; 2, supercooled liquid; 3, glass; 4, crystal

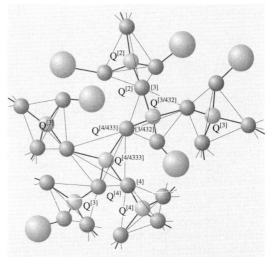

Fig. 3.4-2 Fragment of a sodium silicate glass structure. The SiO_2 tetrahedra are interconnected by the bridging oxygen atoms, thus forming a three-dimensional network. Na_2O units form nonbridging oxygen atoms and act as network modifiers which break the network. The $Q^{[i/klmn]}$ nomenclature describes the connectivity of different atom shells around a selected central atom (after [4.2])

sidered. The most important engineering glasses are mixtures of oxide compounds. For some special requirements, for example a particular optical transmission window or coloration, fluorides, chalcogenides, and colloidal (metal or semiconductor) components are also used.

A very special glass is single-component silica, SiO_2, which is a technological material with extraordinary properties and many important applications.

On a quasi-macroscopic scale (> 100 nm), glasses seem to be homogeneous and isotropic; this means that all structural effects are, by definition, seen only as average properties. This is a consequence of the manufacturing process. If a melt is rapidly cooled down, there is not sufficient time for it to solidify into an ideal, crystalline structure. A structure with a well-defined short-range order (on a scale of less than 0.5 nm, to fulfill the energy-driven bonding requirements of structural elements made up of specific atoms) [4.2] and a highly disturbed long-range order (on a scale of more than 2 nm, disturbed by misconnecting lattice defects and a mixture of different structural elements) (Fig. 3.4-2).

Crystallization is bypassed. We speak of a frozen-in, supercooled, liquid-like structure. This type of quasi-static solid structure is thermodynamically controlled but not in thermal equilibrium and thus is not absolutely stable; it tends to relax and slowly approach an "equilibrium" structure (whatever this may be in a complex multicomponent composition, it represents a minimum of the Gibbs free enthalpy). This also means that all properties change with time and temperature, but in most cases at an extremely low rate which cannot be observed under the conditions of classical applications (in the range of ppm, ppb, or ppt per year at room temperature). However, if the material is exposed to a higher temperature during processing or in the final application, the resulting relaxation may result in unacceptable deformation or internal stresses that then limit its use.

If a glass is reheated, the quasi-solid material softens and transforms into a liquid of medium viscosity in a continuous way. No well-defined melting point exists. The temperature range of softening is called the "transition range". By use of standardized measurement techniques, this imprecise characterization can be replaced by a quasi-materials constant, the "transformation temperature" or "glass temperature" T_g, which depends on the specification of the procedure (Fig. 3.4-3).

As illustrated in Fig. 3.4-4, the continuous variation of viscosity with temperature allows various technologies for hot forming to be applied, for example casting, floating, rolling, blowing, drawing, and pressing, all of which have a specific working point.

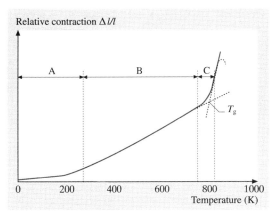

Fig. 3.4-3 Definition of the glass transition temperature T_g by a diagram of length versus temperature during a measurement, as the intersection of the tangents to the elastic region B and to the liquid region to the right of C. In the transformation range C, the glass softens

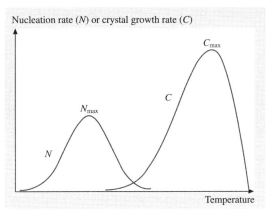

Fig. 3.4-5 Nucleation rate (N) and crystal growth rate (C) of glass as a function of temperature

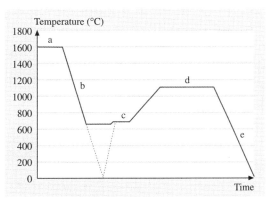

Fig. 3.4-6 Temperature–time schedule for glass-ceramic production: a, melting; b, working; c, nucleation; d, crystallization; e, cooling to room temperature

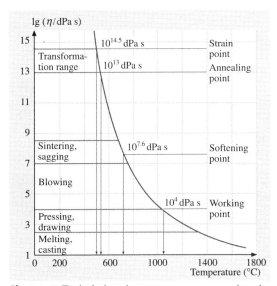

Fig. 3.4-4 Typical viscosity–temperature curve: viscosity ranges for the important processing technologies, and definitions of fixed viscosity points

The liquid-like situation results in thermodynamically induced density and concentration fluctuations, with a tendency toward phase separation and crystallization, which starts via a nucleation step. This can also be used as a technological route to produce unconventional, i.e. inhomogeneous, glasses. Some examples are colored glasses with colloidal inclusions, porous glasses, and glass-ceramics, to name just a few.

The commercially important glass-ceramics consist of a mixture of 30–90% crystallites (< 50 nm in diameter) and a residual glass phase, without any voids. The material is melted and hot-formed as a glass, cooled down, then annealed for nucleation, and finally tempered to allow crystal growth to a percentage that depends on the desired properties [4.3, 4]. Some characteristic features of the manufacturing process are shown in Figs. 3.4-5 and 3.4-6. As the melt is cooled, a region of high velocity of crystal growth has to be passed through, but no nuclei exist in that region. The nuclei are formed at a much lower temperature, where the crystallization is slow again. Thus the material solidifies as a glass in the first process step. In a second process step, the glass is reheated, annealed in the range of maximum nucleation rate N_{max}, and then annealed at a higher temperature in the range of maximum crystal growth rate C_{max}. The overall bulk properties are the averages of the properties of the components. One of the major properties of commercially important glass-ceramics is a near-zero coefficient of thermal expansion.

3.4.1 Properties of Glasses – General Comments

As for all materials, there are many properties for every type of glass described in the literature. In this section only a limited selection can be given. We also restrict the presentation to commercially important glasses and glass-ceramics. For the huge variety of glasses that have been manufactured for scientific purposes only, the original literature must be consulted. Extensive compilations of data of all kind may be found in books, e.g. [4.5–7], and in software packages [4.8, 9], which are based upon these books and/or additional original data from literature, patents, and information from manufacturers worldwide.

Almost all commercially important glasses are silicate-based. For practical reasons, these glasses are subdivided into five major groups which focus on special properties. These groups are:

1. Mass-production glasses, such as window and container glasses, which are soda–lime–silicate glasses. Besides the application-dependant properties such as transparency, chemical resistance, and mechanical strength, the main concern is cost.
2. Technical specialty glasses, such as display or television glasses, glasses for tubes for pharmaceutical packaging, glasses for industrial ware and labware, glasses for metal-to-glass sealing and soldering, and glasses for glassware for consumers. The dominant properties are chemical inertness and corrosion resistance, electrical insulation, mechanical strength, shielding of X-ray or UV radiation, and others. To fulfill these specifications, the glass composition may be complex and even contain rare components.
3. Optical glasses have the greatest variety of chemical components and do not exclude even the most exotic materials, such as rare earth and non-oxide compounds. The dominant properties are of optical origin: refraction and dispersion with extremely high homogeneity, combined with low absorption and light scattering within an extended transmission range, including the infrared and ultraviolet parts of the electromagnetic spectrum. They are produced in comparably small volumes but with raw materials of high purity, and are thus quite expensive.
4. Vitreous silica, as a single-component material, has some extraordinary properties: high transparency from 160 nm to 1800 nm wavelength, high electrical insulation if it is of low hydrogen content, high chemical and corrosion resistance, and quite low thermal expansion with high thermal-shock resistance. A technological handicap is the high glass temperature $T_g \approx 1250\,°C$, depending on the water content.
5. Glass-ceramics are glassy in the first production step. An important property is the final (after ceramization) coefficient of thermal expansion, often in combination with a high elastic modulus and low specific weight. High thermal stability and thermal-shock resistance are a prerequisite for the major applications. This group has the potential for many other applications which require different property combinations [4.3].

This separation into groups seems to be somewhat artificial, in view of the material properties alone, and is justified only by the very different technological conditions used for manufacturing and processing. It corresponds to the specialization of the industry and to a traditional structuring in the literature.

For all these groups, various classes of product defects exist, and these defects may occur in varying concentration. Among these defects may be bubbles, striations, crystalline inclusions, and metal particles, which are relics of a nonideal manufacturing process. There may also be inclusions of foreign components which were introduced with the raw materials as impurities or contamination. If a three-dimensional volume of glass is cooled down, the finite heat conductivity causes the volume elements to have nonidentical histories in time and temperature. As a consequence, the time- and temperature-dependent relaxation processes produce internal mechanical stress, which results in optical birefringence. These "technical" properties are often very important for the suitability of a piece of glass for a specific application, but they are not "intrinsic" properties of the material, and are not considered in this section.

3.4.2 Composition and Properties of Glasses

A very common method for describing the composition of a glass is to quote the content of each component (oxide) by weight fraction (w_i in wt%) or mole fraction (m_i in mol%). If M_i is the molar mass of the component i, the relationship between the two quantities is given by

$$m_i = \frac{100 w_i}{M_i \sum_{j=1}^{n} w_j / M_j}, \quad 1 \leq i \leq n, \quad (4.1)$$

and

$$w_i = \frac{100 m_i M_i}{\sum_{j=1}^{n} m_j M_j}, \quad 1 \leq i \leq n, \quad (4.2)$$

with the side conditions

$$\sum_{i=1}^{n} m_i = 1 \quad \text{and} \quad \sum_{i=1}^{n} w_i = 1. \quad (4.3)$$

If components other than oxides are used, it is advantageous to define fictitious components, for example the component F_2-O (= fluorine–oxygen, with $M = 2 \times 19 - 16 = 22$); this can be used to replace a fluoride with an oxide and this hypothetical oxide component, for example $CaF_2 = CaO + F_2-O$. This reduces the amount of data required

Table 3.4-1 Factors ϱ_{w_i} for the calculation of glass densities [4.10]

Component i	ϱ_{w_i} (g/cm³)
Na$_2$O	3.47
MgO	3.38
CaO	5.0
Al$_2$O$_3$	2.75
SiO$_2$	2.20

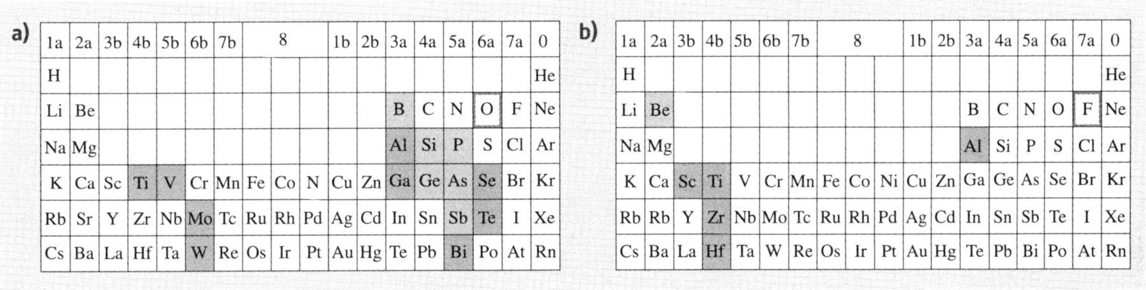

Fig. 3.4-7 (a) Elements whose oxides act as glass-formers (*gray*) and conditional glass-formers (*brown*) in oxide systems (after [4.11]). (b) Elements whose fluorides act as glass-formers (*gray*) and conditional glass-formers (*brown*) in fluoride glass systems (after [4.12])

Table 3.4-2 Composition of selected technical glasses

Glass type	Code[a]	Main components (wt%)					Minor components
		SiO$_2$	B$_2$O$_3$	Al$_2$O$_3$	Na$_2$O	PbO	< 10%
Lead glass	S 8095	57				28	Al$_2$O$_3$, Na$_2$O, K$_2$O
Low dielectric loss glass	S 8248	70	27				Al$_2$O$_3$, Li$_2$O, Na$_2$O, K$_2$O, BaO
Sealing glass	S 8250	69	19				Al$_2$O$_3$, Li$_2$O, Na$_2$O, K$_2$O, ZnO
Duran®	S 8330	80	13				Al$_2$O$_3$, Na$_2$O, K$_2$O
Supremax®	S 8409	52		22			B$_2$O$_3$, CaO, MgO, BaO, P$_2$O$_5$
Sealing glass	S 8465		11	11		75	SiO$_2$
Sealing glass	S 8487	75	16.5				Al$_2$O$_3$, Na$_2$O, K$_2$O
Vycor™	C 7900	96					Al$_2$O$_3$, B$_2$O$_3$, Na$_2$O

[a] Code prefix: S = Schott AG, Mainz; C = Corning, New York.

to handle the different cation–anion combinations (Fig. 3.4-7).

With given weight or mole fractions, a property P can be calculated (or approximated) via a linear relation of the type

$$P = \frac{1}{100}\sum_{i=1}^{n} p_{w_i} w_i \quad \text{or} \quad P = \frac{1}{100}\sum_{i=1}^{n} p_{m_i} m_i \, . \tag{4.4}$$

The component-specific factors p_i were first systematically determined by *Winckelmann* and *Schott* [4.13], and are listed in detail in the standard literature, e.g. [4.14].

As an example, the density ϱ can be calculated via (4.4) from: $P = 1/\varrho$, $p_{w_i} = 1/\varrho_{w_i}$, and the data in Table 3.4-1.

The composition of selected technical glasses is given in Table 3.4-2.

3.4.3 Flat Glass and Hollowware

By far the highest glass volume produced is for windows and containers. The base glass is a soda–lime composition, which may be modified for special applications. Owing to the Fe content of the natural raw materials, a green tint is observed for thicknesses that are not too small.

If the product is designed to have a brown or green color (obtained by the use of reducing or oxidizing melting conditions), up to 80% waste glass can be used as a raw material.

3.4.3.1 Flat Glass

Plate glass for windows is mainly produced by floating the melt on a bath of molten tin. All manufacturers worldwide use a soda–lime-type glass, with the average composition given in Table 3.4-3. For some other applications, this base glass may be modified by added coloring agents, for example Cr and Fe oxides.

For architectural applications, the surface(s) may be modified by coatings or grinding to achieve specific

Table 3.4-3 Average composition in wt% and viscosity data, for soda–lime glass, container glass, and Borofloat glass

	Soda–lime glass	Container glass	Borofloat® 33
Composition (wt%)			
SiO_2	71–73	71–75	81
$CaO + MgO$	9.5–13.5	10–15	
$Na_2O + K_2O$	13–16	12–16	4
B_2O_3	0–1.5		13
Al_2O_3	0.5–3.5		2
Other	0–3	0–3	
Temperature (°C)			
at viscosity of $10^{14.5}$ dPa s			518
at viscosity of 10^{13} dPa s	525–545		560
at viscosity of $10^{7.6}$ dPa s	717–735		820
at viscosity of 10^4 dPa s	1015–1045		1270
Density (g/cm³)	2.5		2.2
Young's modulus (GPa)	72		64
Poisson's ratio			0.2
Knoop hardness			480
Bending strength (MPa)	30		25
Thermal expansion coefficient α, 20–300 °C (10^{-6}/K)	9.0		3.25
Specific heat capacity c_p, 20–100 °C (kJ/kg K)			0.83

Table 3.4-3 Average composition in wt% and viscosity data for soda–lime glass, cont.

	Soda–lime glass	Container glass	Borofloat®33
Heat conductivity λ at 90 °C (W/m kg)			1.2
Chemical resistance			
Water			HGB1/HGA1
Acid			1
Alkali			A2
Refractive index n_d			1.47140
Abbe value ν_e			65.41
Stress-optical constant K (10^{-6} mm^2/N)			4.0
Dielectric constant ε_r			4.6
Dielectric loss tan δ at 25 °C, 1 MHz			37×10^{-4}
Volume resistivity ρ			
at 250 °C (Ω cm)			10^8
at 350 °C (Ω cm)			$10^{6.5}$

optical effects, such as decorative effects, a higher reflectivity in certain spectral regions (e.g. the IR), or an opaque but translucent appearance. Some other applications require antireflection coatings.

If steel wires are embedded as a web into rolled glass, the resulting product can be used as a window glass for areas where improved break-in protection is required.

Fire-protecting glasses are designed to withstand open fire and smoke for well-defined temperature–time programs, for example from 30 min to 180 min with a maximum temperature of more than 1000 °C (test according to DIN 4102, Part 13). The main goal is to avoid the glass breaking under thermal load. There are various ways to achieve this, such as mechanical wire reinforcement, prestressing the plate by thermal or surface-chemical means, or reducing the thermal expansion by using other glass compositions, for example the floated borosilicate glass Pyran® (Schott).

In automotive applications, flat glass is used for windows and mirrors. Single-pane safety glass is produced by generating a compressive stress near to the surface by a special tempering program: a stress of 80–120 MPa is usual. In the case of a breakage the window breaks into small pieces, which greatly reduces the danger of injuries.

Windshields are preferably made of laminated compound glass. Here, two (or more) layers are joined under pressure with a tough plastic foil.

The compound technique is also used for armor-plate glass: here, at least four laminated glass layers with a total thickness of at least 60 mm are used.

For plasma and LCD displays and photovoltaic substrates, extremely smooth, planar, and thin sheets of glass are needed which fit the thermal expansion of the electronic materials in direct contact with them. These panes must remain stable in geometry (no shrinkage) when they are processed to obtain the final product.

3.4.3.2 Container Glass

For packaging, transportation, storage of liquids, chemicals, pharmaceuticals, etc., a high variety of hollowware is produced in the shape of bottles and tubes. Most of the products are made from soda–lime glass (Table 3.4-3) directly from the glass melt. Vials and ampoules are manufactured from tubes; for pharmaceuticals these are made from chemically resistant borosilicate glass.

Clear glass products have to be made from relatively pure raw materials. To protect the contents from light, especially UV radiation, the glass may be colored brown or green by the addition of Fe or Cr compounds. For other colors, see Sect. 3.4.8.3.

3.4.4 Technical Specialty Glasses

Technical glasses are special glasses manufactured in the form of tubes, rods, hollow vessels, and a variety of special shapes, as well as flat glass and granular form. Their main uses are in chemistry, pharmaceutical packaging, electronics, and household appliances. A list of typical applications of different glass types is given in Table 3.4-10, and a list of their properties is given in Table 3.4-11.

The multitude of technical glasses can be roughly arranged in the following four groups, according to their oxide composition (in weight percent). However, certain glass types fall between these groups or even completely outside.

Borosilicate Glasses. The network formers are substantial amounts of silica (SiO_2) and boric oxide ($B_2O_3 > 8\%$). The following subtypes can be differentiated:

- *Non-alkaline-earth borosilicate glass*: with $SiO_2 > 80\%$ and B_2O_3 at $12-13\%$, these glasses show high chemical durability and low thermal expansion (about $3.3 \times 10^{-6}\,K^{-1}$).
- *Alkaline-earth-containing glasses*: with $SiO_2 \approx 75\%$, B_2O_3 at $8-12\%$, and alumina (Al_2O_3) and alkaline earths up to 5%, these glasses are softer in terms of viscosity, and have high chemical durability and a thermal expansion in the range of $4.0-5.0 \times 10^{-6}\,K^{-1}$.
- *High-borate glasses*: with $65-70\%\ SiO_2$, $15-25\%\ B_2O_3$, and smaller amounts of Al_2O_3 and alkalis, these glasses have a low softening point and a thermal expansion which is suitable for glass-to-metal seals.

Alkaline-Earth Aluminosilicate Glasses. These glasses are free from alkali oxides, and contain $52-60\%\ SiO_2$, $15-25\%\ Al_2O_3$, and about 15% alkaline earths. A typical feature is a very high transformation temperature.

Alkali–Lead Silicate Glasses. These glasses contain over 10% lead oxide (PbO). Glasses with $20-30\%$ PbO, $54-58\%\ SiO_2$, and about 14% alkalis are highly insulating and provide good X-ray shielding. They are used in cathode-ray tube components.

Alkali–Alkaline-Earth Silicate Glasses (Soda–Lime Glasses). This is the oldest glass type, which is produced in large batches for windows and containers (see Sect. 3.4.3). Such glasses contain about $71\%\ SiO_2$, $13-16\%$ alkaline earths $CaO + MgO$, about 15% alkali (usually Na_2O), and $0-2\%\ Al_2O_3$.

Variants of the basic composition can contain significant amounts of BaO with reduced alkali and alkaline-earth oxides: these are used for X-ray shielding of cathode-ray tube screens.

3.4.4.1 Chemical Stability of Glasses

Characteristically, glasses are highly resistant to water, salt solutions, acids, and organic substances. In this respect, they are superior to most metals and plastics. Glasses are attacked to a significant degree only by hydrofluoric acid, strongly alkaline solutions, and concentrated phosphoric acid; this occurs particularly at higher temperatures.

Chemical reactions with glass surfaces, induced by exchange, erosion, or adsorption processes, can cause the most diverse effects, ranging from virtually invisible surface modifications to opacity, staining, thin films with interference colors, crystallization, holes, and rough or smooth ablation, to name but a few such effects. These changes are often limited to the glass surface, but in extreme cases they can completely destroy or dissolve the glass. The glass composition, stress medium, and operating conditions will decide to what extent such chemical attacks are technically significant.

Chemical Reaction Mechanisms with Water, Acid, and Alkaline Solutions

Chemical stability is understood as the resistance of a glass surface to chemical attack by defined agents; here, temperature, exposure time, and the condition of the glass surface play important roles.

Every chemical attack on glass involves water or one of its dissociation products, i.e. H^+ or OH^- ions. For this reason, we differentiate between hydrolytic (water), acid, and alkali resistance. In water or acid attack, small amounts of (mostly monovalent or divalent) cations are leached out. In resistant glasses, a very thin layer of silica gel then forms on the surface, which normally inhibits further attack (Fig. 3.4-8a,b). Hydrofluoric acid, alkaline solutions, and in some cases phosphoric acid, however, gradually destroy the silica framework and thus ablate the glass surface in total (Fig. 3.4-8c). In contrast, water-free (i.e. organic) solutions do not react with glass.

Chemical reactions are often increased or decreased by the presence of other components. Alkali attack on glass is thus hindered by certain ions, particularly those

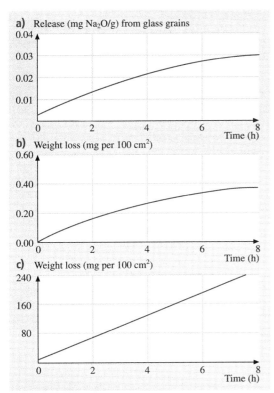

Fig. 3.4-8a–c Attack by (a) water, (b) acid, and (c) alkaline solution on chemically resistant glass as a function of time

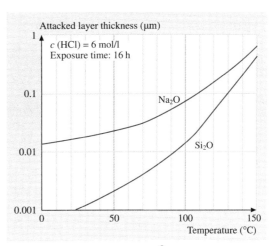

Fig. 3.4-9 Acid attack on Duran® 8330 as a function of temperature, determined from leached amounts of Na_2O and SiO_2

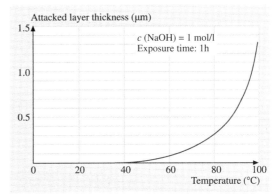

Fig. 3.4-10 Alkali attack on Duran® 8330 as a function of temperature, determined from weight loss

of aluminium. On the other hand, complex-forming compounds such as EDTA, tartaric acid, and citric acid can increase the solubility. In general terms, the glass surface reacts with solutions which induce small-scale exchange reactions and/or adsorption. Such phenomena are observed, for example, in high-vacuum technology when residual gases are removed, and in certain inorganic-chemical operations, when small amounts of adsorbed chromium, resulting from treatment with chromic acid, are removed.

Because acid and alkali attacks on glass are fundamentally different, silica-gel layers produced by acid attack obviously are not necessarily effective against alkali solutions and may be destroyed. Conversely, the presence of ions that inhibit alkali attack does not necessarily represent protection against acids and water. The most severe chemical exposure is therefore an alternating treatment with acids and alkaline solutions. As in all chemical reactions, the intensity of the interaction increases rapidly with increasing temperature (Figs. 3.4-9 and 3.4-10).

In the case of truly ablative solutions such as hydrofluoric acid, alkaline solutions, and hot concentrated phosphoric acid, the rate of attack increases rapidly with increasing concentration (Fig. 3.4-11). As can be seen in Fig. 3.4-12, this is not true for the other frequently used acids.

Determination of the Chemical Stability

In most cases, either is the glass surface analyzed in its "as delivered" condition (with the original fire-polished surface), or the basic material is analyzed with its fire-

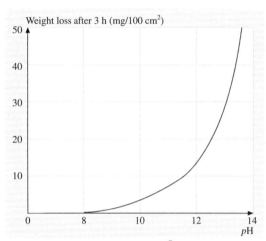

Fig. 3.4-11 Alkali attack on Duran® 8330 at 100 °C as a function of pH value

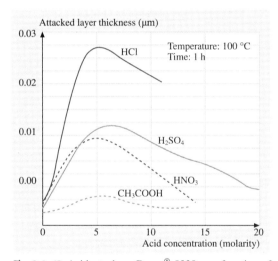

Fig. 3.4-12 Acid attack on Duran® 8330 as a function of concentration

polished surface removed by mechanical or chemical ablation, or after crushing.

The standardized DIN (Deutsches Institut für Normung, German Institute for Standardization) test methods, which are universally and easily applicable, are the most reliable analysis methods. They include the determination of hydrolytic resistance (by two grain-titration methods and one surface method), of acid resistance to hydrochloric acid, and of alkali resistance to a mixture of alkaline solutions. Details are described in [4.2, 15] and summarized in Tables 3.4-4–3.4-6. The DIN classes of hydrolytic, acid, and alkali resistance of technical glasses are also listed Table 3.4-11, second page, last three colums.

Significance of the Chemical Stability

Release of Glass Constituents. In various processes in chemical technology, pharmaceutical manufacture, and laboratory work, the glass material used is expected to release no constituents (or a very minimal amnount) into reacting solutions or stored specimens.

Because even highly resistant materials such as non-alkaline-earth and alkaline-earth borosilicate glasses do react to a very small degree with the surrounding media, the fulfillment of this requirement is a question of quantity and of detection limits. Concentrations of 10^{-6}–10^{-9} (i. e. trace amounts), which are measurable today with highly sophisticated analytical instruments, can be released even from borosilicate glasses in the form of SiO_2, B_2O_3, and Na_2O, depending on the conditions. However, solutions in contact with high-grade colorless Duran® laboratory glass will not be contaminated by Fe, Cr, Mn, Zn, Pb, or other heavy-metal ions.

Undesirable Glass Surface Modifications. When an appreciable interaction between a glass surface and an aqueous solution occurs, there is an ion exchange in which the easily soluble glass components are replaced by H^+ or OH^- ions. This depletion of certain glass components in the surface leads to a corresponding enrichment in silica, which is poorly soluble, and thus to the formation of a silica-gel layer. This layer proves, in most cases, to be more resistant than the base glass. When its thickness exceeds about 0.1–0.2 μm, interference colors caused by the different refractive indices of the layer and the base glass make this silica-gel layer visible to the unaided eye. With increasing layer thickness it becomes opaque and finally peels off, destroying the glass. Between these stages there is a wide range of possible surface modifications, some of which, although optically visible, are of no practical significance, whereas others must be considered.

In the case of less resistant glasses, small amounts of water (from air humidity and condensation) in the presence of other agents such as carbon dioxide or sulfur oxides can lead to surface damage. In the case of sensitive glasses, hand perspiration or impurities left by detergents can sometimes induce strongly adhering surface defects, mostly recognizable as stains. If a contaminated glass surface is reheated (> 350–400 °C), the

Table 3.4-4 Hydrolytic classes according to DIN 12111 (ISO 719)

Hydrolytic class	Acid consumption of 0.01 mol/l hydrolytic acid per g glass grains (ml/g)	Base equivalent as Na_2O per g glass grains (µg/g)	Possible designation
1	Up to 0.10	Up to 31	Very high resistance
2	Above 0.10, up to 0.20	Above 31, up to 62	High resistance
3	Above 0.20, up to 0.85	Above 62, up to 264	Medium resistance
4	Above 0.85, up to 2.0	Above 264, up to 620	Low resistance
5	Above 2.0, up to 3.5	Above 620, up to 1085	Very low resistance

Table 3.4-5 Acid classes according to DIN 12116

Acid class	Designation	Half loss in weight after 6 h (mg/100 cm^2)
1	High acid resistance	Up to 0.7
2	Good acid resistance	Above 0.7, up to 1.5
3	Medium acid attack	Above 1.5, up to 15
4	High acid attack	Above 15

Table 3.4-6 Alkali classes according to DIN ISO 695

Alkali class	Designation	Loss in weight after 3 h (mg/100 cm^2)
1	Low alkali attack	Up to 75
2	Medium alkali attack	Above 75, up to 175
3	High alkali attack	Above 175

contaminants or some of their components may burn in. Normal cleaning processes will then be ineffective and the whole surface layer has to be removed (e.g. by etching).

Desirable Chemical Reactions with the Glass Surface (Cleaning and Etching). Very strong reactions between aqueous agents and glass can be used for the thorough cleaning of glass. The complete ablation of a glass layer leads to the formation of a new surface.

Hydrofluoric acid reacts most strongly with glass. Because it forms poorly soluble fluorides with a great number of glass constituents, it is mostly used only in diluted form. The best etching effect is usually achieved when another acid (e.g. hydrochloric or nitric acid) is added. A mixture of seven parts by volume of water, two parts of concentrated hydrochloric acid ($c = 38\%$) and one part of hydrofluoric acid ($c = 40\%$) is recommended for a moderate surface ablation of highly resistant borosilicate glasses. When chemically less resistant glasses (e.g. Schott 8245 and 8250) are exposed for 5 min to a stirred solution at room temperature, a surface layer with a thickness of $1-10\,\mu m$ is ablated, and a transparent, smooth, completely new surface is produced.

Glasses can also be ablated with alkaline solutions, but the alkaline etching process is much less effective.

3.4.4.2 Mechanical and Thermal Properties

Viscosity

As described earlier, the viscosity of glasses increases by 15–20 orders of magnitude during cooling. Within this viscosity range, glasses are subject to three different thermodynamic states:

- the melting range, above the liquidus temperature T_s;
- the range of the supercooled melt, between the liquidus temperature T_s and the transformation temperature T_g, which is defined by ISO 7884-8;
- the frozen-in, quasi-solid melt range ("glass range"), below the transformation temperature T_g.

The absence of any significant crystallization in the range of the supercooled melt (see Fig. 3.4-1, line segment 2) is of the utmost importance for glass formation. Hence a basically steady, smooth variation in the viscosity in all temperature regions is a fundamental characteristic of glasses (Fig. 3.4-4) and a crucial property for glass production. Figure 3.4-13 shows the strongly differing temperature dependences of the viscosity for some glasses. The best mathematical expression for practical purposes is the VFT (Vogel, Fulcher, and Tammann) equation,

$$\log \eta(T) = A + B/(T - T_0)\,, \qquad (4.5)$$

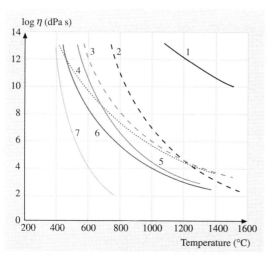

Fig. 3.4-13 Viscosity–temperature curves for some important technical glasses. 1, fused silica; 2, 8405; 3, 8330; 4, 8248; 5, 8350; 6, 8095; 7, 8465. Glasses with steep gradients (such as 7) are called "short" glasses, and those with relatively shallow gradients (such as 4) are called "long" glasses

where A, B, and T_0 are glass-specific constants (Table 3.4-7).

Somewhat above 10^{10} dPa s, the viscosity becomes increasingly time-dependent. With increasing viscosity (i. e. with decreasing temperature), the delay in establishing structural equilibrium finally becomes so large that, under normal cooling conditions, the glass structure at 10^{13} dPa s can be described as solidified or "frozen-in". This temperature (for which a method of measurement is specified by ISO 7884-4) is called the "annealing point". At this viscosity, internal stresses in the glass are released after ≈ 15 min annealing time, while the dimensional stability of the glass is sufficient for many purposes, and its brittleness (susceptibility to cracking) is almost fully developed.

The lower limit of the annealing range is indicated by the "strain point", at which the glass has a viscosity of $10^{14.5}$ dPa s (determined by extrapolation from the viscosity–temperature curve). For most glasses, the strain point lies about 30–40 K below the annealing point. Relaxation of internal stresses here takes 3–5 h. The strain point marks the maximum value for short-term heat load. Thermally prestressed glasses, in contrast, show significant stress relaxation even at 200–300 K below T_g. For glass objects with precisely defined dimensions (e.g. etalons and gauge blocks) and in the case of extreme demands on the stability of certain properties of the glass, application temperatures of 100–200 °C can be the upper limit.

Strength

The high structural (theoretical) strength of glasses and glass-ceramics ($> 10^4$ N/mm^2 = 10 GPa) is without practical significance, because the strength of glass articles is determined by surface defects induced by wear, such as tiny chips and cracks (Griffith flaws), at whose tips critical stress concentrations may be induced by a mechanical load, especially if the load is applied perpendicular to the plane of the flaw (fracture mode I). Glasses and glass-ceramics, in contrast to ductile materials such as metals, show no plastic flow and behave under a tensile stress σ in as brittle a manner as ceramics. A flaw will result in a fracture if the "stress intensity factor"

$$K_I = 2\sigma\sqrt{a} > K_{Ic}, \qquad (4.6)$$

where a is the depth of the flaw and K_{Ic} is the "critical stress intensity factor", a material constant which is temperature- and humidity-dependent: see Table 3.4-8.

For $K_{Ic} = 1$ MPa $\sqrt{m}$ and a stress $\sigma = 50$ MPa, the critical flaw depth a_c is 100 µm. Thus very small flaws

Table 3.4-7 Parameters of the VFT equation (4.5) for the glasses in Fig. 3.4-13

Glass	A	B (°C)	T_0 (°C)
8095	−1.5384	4920.68	96.54
8248	−0.2453	4810.78	126.79
8330	−1.8500	6756.75	105.00
8350	−1.5401	4523.85	218.10
8405	−2.3000	5890.50	65.00
8465	−1.6250	1873.73	256.88
Fused silica	−7.9250	31 282.9	−415.00
Soda–lime glass	−1.97	4912.5	475.4

can cause cracking at a comparatively low stress level, and the practical strength of a glass is not a materials constant!

Surface Condition. As a result of wear-induced surface defects, glass and glass-ceramic articles have practical tensile strengths of $20-200\,\text{N/mm}^2 = 20-200\,\text{MPa}$, depending on the surface condition and the atmospheric-exposure condition. To characterize the strength, a Weibull distribution for the cumulative failure probability F is assumed:

$$F(\sigma) = 1 - \exp\left[-(\sigma/\sigma_c)^m\right]\,, \qquad (4.7)$$

where $0 \leq F(\sigma) \leq 1$ is the probability of a fracture if the applied stress is less than σ; σ_c denotes the characteristic value (approximately the mean value of the distribution), and m is the Weibull modulus of the distribution (which determines the standard deviation). To obtain reproducible measurements, the surface is predamaged by grinding with a narrow grain-size distribution (Fig. 3.4-14).

Only a slight – as a rule with neglible – dependence on the chemical composition is found for silicate glasses (Table 3.4-8).

Stress Rate. The rate of increase of the stress and the size of the glass area exposed to the maximum stress have to be considered for the specification of a strength value. In contrast to the rapid stress increase occurring in an impact, for example, a slowly increasing tensile stress or continuous stress above a certain critical limit may – as a result of stress corrosion cracking – cause the propagation of critical surface flaws and cracks and thus enhance their effect. Hence the tensile strength is time- and stress-rate-dependent (this is mainly important for test loads), as shown in Fig. 3.4-15. Independent of surface damage or the initial tensile strength, increasing

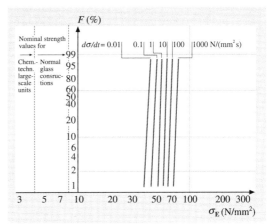

Fig. 3.4-15 Failure probability F of a predamaged surface ($100\,\text{mm}^2$; grain size 600) for various rates of increase of stress $d\sigma/dt$. A, range of nominal strength for large-scale units in chemical technology; B, range of nominal strength for normal glass structures

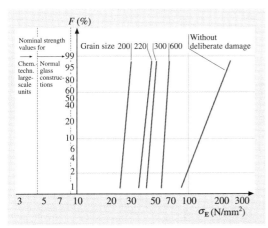

Fig. 3.4-14 Failure probability F for samples abraded by variously sized grains. Predamaged surface area $100\,\text{mm}^2$, rate of stress increase $d\sigma/dt = 10\,\text{MPa s}^{-1}$. A, range of nominal strength for large-scale units in chemical technology; B, range of nominal strength for normal glass structures

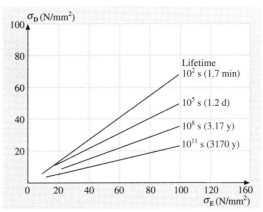

Fig. 3.4-16 Time-related strength σ_D (strength under constant loading) of soda–lime glass compared with the experimental strength σ_E at $d\sigma/dt = 10\,\text{N/mm}^2\,\text{s}$ for various lifetimes, in a normal humid atmosphere

the rate of increase of the stress by a factor of 10 results in an increase in the strength level of about 15%.

Constant Loading. Fracture analysis of the effect and behavior of cracks in glasses and glass-ceramics yield further information about the relationship between the experimentally determined tensile strength σ_E (usually measured for a rapidly increasing load) and the tensile strength σ_D expected under constant loading (= fatigue strength), as shown in Fig. 3.4-16. Such analyses show that, depending on the glass type, the tensile strength under constant loading σ_D (for years of loading) may amount to only about 1/2 to 1/3 of the experimental tensile strength σ_E.

Area Dependence. The larger the stressed area, the higher is the probability of large defects (large crack depths) within this area. This relationship is important for the transfer of experimental tensile strengths, which are mostly determined with relatively small test samples, to practical glass applications such as pipelines, where many square meters of glass can be uniformly stressed (Fig. 3.4-17).

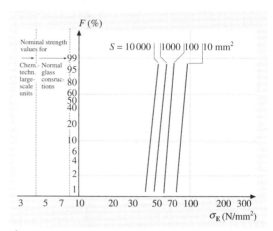

Fig. 3.4-17 Failure probability F for differently sized stressed areas S. All samples abraded with 600 mesh grit, stress rate $d\sigma/dt = 10\,\text{N/mm}^2$ s. A, range of nominal strength for large-scale units in chemical technology; B, range of nominal strength for normal glass structures

Elasticity

The ideal brittleness of glasses and glass-ceramics is matched by an equally ideal elastic behavior up to breaking point. The elastic moduli for most technical glasses lie within a range of 50–90 kN/mm². The mean value of 70 kN/mm² is about equal to the Young's modulus of aluminium (see Table 3.4-11, first page, column 7).

Coefficient of Linear Thermal Expansion

With few exceptions, the length and the volume of glasses increase with increasing temperature (positive coefficient).

The typical curve begins with a zero gradient at absolute zero (Fig. 3.4-3) and increases slowly. At about room temperature (section A in Fig. 3.4-3), the curve shows a distinct bend and then gradually increases (section B, the quasi-linear region) up to the beginning of the experimentally detectable plastic behavior. Another distinct bend in the expansion curve characterizes the transition from a predominantly elastic to a more plastic behavior of the glass (section C, the transformation range). As a result of increasing structural mobility, the temperature dependence of almost all glass properties changes distinctly in this range. Figure 3.4-18 shows the linear thermal expansion curves of five glasses; 8330 and 4210 roughly define the normal range of technical glasses, with expansion coefficients

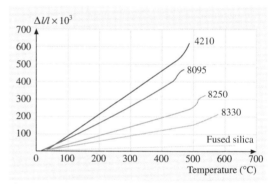

Fig. 3.4-18 Linear thermal expansion coefficients of various technical glasses and of fused silica

$\alpha_{(20\,°C/300\,°C)} = 3.3\text{–}12.0 \times 10^{-6}\,/\text{K}$ (see Table 3.4-11, first page, third column).

The linear thermal expansion is an essential variable in determining the sealability of glasses to other materials and in determining thermally induced stress formation, and is therefore of prime importance for applications of glasses.

Thermal Stresses. Owing to the low thermal conductivity of glasses (typically 0.9–1.2 W/m K at 90 °C, and a minimum of 0.6 W/m K for high-lead-content glasses), temperature changes produce relatively high temperature differences ΔT between the surface and the interior, which, depending on the elastic properties E (Young's modulus) and μ (Poisson's ratio) and on the coefficient of linear thermal expansion α, can result in stresses

$$\sigma = \frac{\Delta T \alpha E}{(1-\mu)} \quad \text{N/mm}^2 \,. \tag{4.8}$$

In addition to the geometric factors (shape and wall thickness), the material properties α, E, and μ decisively influence the thermal strength of glasses subjected to temperature variations and/or thermal shock. Thermal loads in similar articles made from different glasses are easily compared by means of the characteristic material value

$$\varphi = \frac{\sigma}{\Delta T} = \frac{\alpha E}{(1-\mu)} \quad \text{N/(mm}^2 \text{ K)} \,, \tag{4.9}$$

which indicates the maximum thermally induced stress to be expected in a flexure-resistant piece of glass for a local temperature difference of 1 K. Because cracking originates almost exclusively from the glass surface and is caused there by tensile stress alone, cooling processes are usually much more critical than the continuous rapid heating of glass articles.

3.4.4.3 Electrical Properties

Glasses are used as electrically highly insulating materials in electrical engineering and electronics, in the production of high-vacuum tubes, lamps, electrode seals, hermetically encapsulated components, high-voltage insulators, etc. Moreover, glasses may be used as insulating substrates for electrically conducting surface layers (in surface heating elements and data displays).

Volume Resistivity

Electrical conductivity in technical silicate glasses is, in general, a result of the migration of ions – mostly alkali ions. At room temperature, the mobility of these ions is usually so small that the volume resistivity, with values above $10^{15}\,\Omega\,\text{cm}$, is beyond the range of measurement. The ion mobility increases with increasing temperature. Besides the number and nature of the charge carriers, the structural effects of other components also influence the volume resistivity and its relationship to temperature. The Rasch and Hinrichsen law applies to this relationship at temperatures below the transformation range:

$$\log \rho = A - B/T \,, \tag{4.10}$$

where ρ is the electrical volume resistivity in $\Omega\,\text{cm}$, A, B are constants specific to the particular glass, and T is the absolute temperature in K.

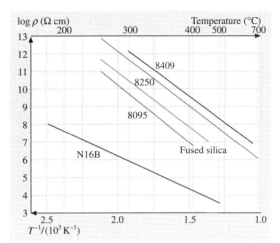

Fig. 3.4-19 Electrical volume resistivity of various technical glasses and fused silica as a function of reciprocal absolute temperature

A plot of $\log \rho = f(1/T)$ thus yields straight lines (Fig. 3.4-19). Because of the relatively small differences in slope for most glasses, the electrical insulation of glasses is often defined only by the temperature at which the resistivity is $10^8\,\Omega\,\text{cm}$. According to DIN 52326, this temperature is denoted by T_{k100}. The international convention is to quote volume resistivities at 250 °C and 350 °C (Table 3.4-11, second page, second column), from which the constants A and B

Table 3.4-8 Fracture toughness of some glasses

Glass	K_{Ic} (MPa $\sqrt{\text{m}}$)
BK7	1.08
F5	0.86
SF6	0.74
K50	0.77
Duran®	0.85

Table 3.4-9 Parameters of the volume resistivity (4.10) of the glasses in Fig. 3.4-19

Glass	A	B (K)
8095	− 2.863	− 6520.0
8250	− 0.594	− 5542.0
8409	− 0.463	− 6520.0
N16B	− 1.457	− 3832.4
Fused silica	− 0.394	− 6222.4

(Table 3.4-9) and various other values below T_g can be calculated.

Surface Resistivity

The generally very high volume resistivity of glasses at room temperature has superimposed on it in a normal atmosphere a surface resistivity which is several orders of magnitude lower. The all-important factor is the adsorption of water on the glass surface. Depending on the glass composition, surface resistivities of $10^{13}-10^{15}$ Ω occur at low relative humidities, and 10^8-10^{10} Ω at high relative humidities. Above 100 °C, the effect of this hydrated layer disappears almost completely. (Treatment with silicones also considerably reduces this effect.)

Dielectric Properties

With dielectric constants generally between 4.5 and 8, technical glasses behave like other electrically insulating materials. The highest values are obtained for lead glasses such as 8531 ($\varepsilon_r = 9.5$) and for ultra-high-lead-content solder glasses ($\varepsilon_r \sim 20$). The dependence of the dielectric constants ε_r on frequency and temperature is relatively small (Fig. 3.4-20). For a frequency range of $50-10^9$ Hz, ε_r values generally do not vary by more than 10%.

The dielectric dissipation factor $\tan\delta$ is frequency- and temperature-dependent. Owing to the diverse mechanisms which cause dielectric losses in glasses, there is a minimum of $\tan\delta$ in the region of 10^6-10^8 Hz, and increasing values at lower and higher frequencies (Fig. 3.4-21).

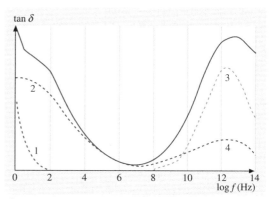

Fig. 3.4-21 Schematic representation of the frequency spectrum of dielectric losses in glasses at room temperature. The *solid curve* gives the total losses, made up of (1) conduction loss, (2) relaxation loss, (3) vibration loss, and (4) deformation loss

At 10^6 Hz, the dissipation factors $\tan\delta$ for most glasses lie between 10^{-2} and 10^{-3}; fused silica, with a value of 10^{-5}, has the lowest dissipation factor of all glasses. The special glass 8248 has relatively low losses, and in this cases $\tan\delta$ increases only slightly up to 5.5 GHz (where $\tan\delta = 3 \times 10^{-3}$).

The steep increase in dielectric losses with increasing temperature (Fig. 3.4-22) can lead to instability, i.e. overheating of the glass due to dielectric loss energy in the case of restricted heat dissipation and corresponding electrical power.

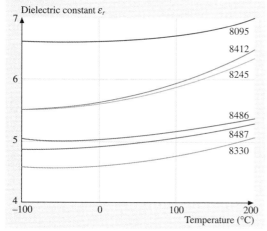

Fig. 3.4-20 Dielectric constant ε_r of electrotechnical glasses as a function of temperature, measured at 1 MHz

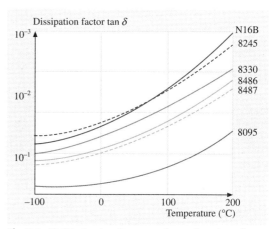

Fig. 3.4-22 Dissipation factor $\tan\delta$ as a function of temperature in the range $-100\,°C < T < +200\,°C$, measured at 1 MHz

Dielectric Strength

Some approximate values for the dielectric strength of glasses are a field strength of 20–40 kV/mm for a glass thickness of 1 mm at 50 Hz at 20 °C, and 10–20 kV/mm for greater thicknesses. At higher temperatures and frequencies, decreasing values can be expected.

3.4.4.4 Optical Properties

Refraction of Light

The refractive indices n_d of technical glasses at a wavelength of $\lambda_d = 587.6$ nm generally lie within the range 1.47–1.57. The exceptions to this rule are lead glasses with PbO contents of over 35% (e.g. glass 8531, which has $n_d = 1.7$; see Table 3.4-11, second page, fourth column). The principal dispersion $n_F - n_C$ ($\lambda_F = 486.1$ nm, $\lambda_C = 656.3$ nm) of technical glasses lies between 0.007 and 0.013.

At perpendicular incidence, the reflectance R_d of a glass–air interface is 3.6% to 4.9%.

The transmittance τ_d and the reflectance ρ_d of a non-absorbing, planar, parallel-sided glass plate with two glass–air interfaces, with multiple reflections taken into account, can be calculated from the refractive index as

$$\tau_d = \frac{2n_d}{n_d^2 + 1} \qquad (4.11)$$

and

$$\rho_d = \frac{(n_d - 1)^2}{n_d + 1}. \qquad (4.12)$$

The transmittance τ_d at perpendicular incidence has values between 90.6% and 93.1%.

Stress Birefringence

Owing to its structure, glass is an isotropic material. Mechanical stress causes anisotropy, which manifests itself as stress-induced birefringence. A light beam, after passing through a plate of thickness d which is subjected to a principal-stress difference $\Delta\sigma$, shows an optical path difference Δs between the two relevant polarization directions. This path difference can either be estimated by means of the birefringence colors or be measured with a compensator, and is given by

$$\Delta s = K d \Delta \sigma \text{ nm}, \qquad (4.13)$$

where K is the stress-optical coefficient of the glass (determined according to DIN 52314),

$$K = \frac{\Delta s}{d} \frac{1}{\Delta\sigma} \text{ mm}^2/\text{N}. \qquad (4.14)$$

Many glasses have stress-optical coefficients of about 3×10^{-6} mm²/N, and borosilicate glasses have values

of up to 4×10^{-6} mm²/N. High-lead-content glasses can have values down to nil or even negative (Table 3.4-11, second page, fifth column).

Light Transmittance

The transmittance due to the refractive index can be further reduced by coloring agents (oxides of transition elements or colloids) or by fine particles in the glass which have a different refractive index (in this case light scattering occurs, giving opal glasses).

Absorption caused by impurities such as Fe_2O_3 and by some major glass components such as PbO strongly reduces transparency in the UV range. Particularly good UV-transmitting multicomponent glasses have a cutoff

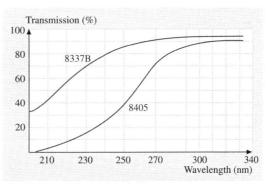

Fig. 3.4-23 UV transmission of highly UV-transparent glasses 8337B and 8405 for 1 mm glass thickness

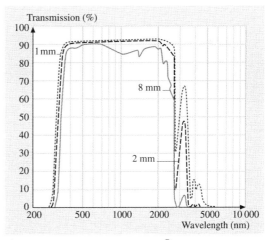

Fig. 3.4-24 Transmission of Duran® 8330 for thicknesses of 1, 2, and 8 mm

(50% value) at a wavelength of 220 nm (Fig. 3.4-23); normal technical glasses already absorb considerably at 300 nm.

In the IR range, absorption caused by impurities such as H_2O and by lattice vibrations limits the transmittance (Fig. 3.4-24).

Table 3.4-10 Schott technical specialty glasses and their typical applications

8095	Lead glass (28% PbO), electrically highly insulating, for general electrotechnical applications
8245	Sealing glass for Fe–Ni–Co alloys and molybdenum, minimum X-ray absorption, chemically highly resistant
8248	Borosilicate glass (of high B_2O_3 content), minimum dielectric losses up to the GHz range, electrically highly insulating
8250	Sealing glass for Ni–Fe–Co alloys and molybdenum, electrically highly insulating
8252	Alkaline-earth aluminosilicate glass for high-temperature applications, for sealing to molybdenum
8253	Alkaline-earth aluminosilicate glass for high-temperature applications, for sealing to molybdenum
8321	Alumino-borosilicate glass for TFT displays
8326	SBW glass, chemically highly resistant
8330	Duran®, borosilicate glass, general-purpose glass for apparatus for the chemical industry, pipelines, and laboratory glassware
8337B	Borosilicate glass, highly UV-transmitting, for sealing to glasses and to metals of the Kovar and Vacon-10 ranges and tungsten
8350	AR glass®, soda–lime silicate glass tubing
8405	Highly UV-transmitting soft glass
8409	Supremax® (black identification line), alkali-free, for high application temperatures in thermometry, apparatus construction, and electrical engineering
8412 [a]	Fiolax®, clear (blue identification line), neutral, glass tubing (chemically highly resistant) for pharmaceutical packaging
8414	Fiolax®, amber (blue identification line), neutral, glass tubing (chemically highly resistant) for pharmaceutical packaging
8415	Illax®, amber tubing glass for pharmaceutical packaging
8421	Sealing glass for seals to NiFe45 (DIN 17745) and compression seals
8422	Sealing glass for seals to NiFe47 or 49 (DIN 17745) and compression seals
8436	Particularly resistant to sodium vapor and alkaline solutions, suitable for sealing to sapphire
8486	Suprax®, borosilicate glass, chemically and thermally resistant, suitable for sealing to tungsten
8487	Sealing glass for tungsten, softer than 8486
8488	Borosilicate glass, chemically and thermally resistant
8490	Black glass, light-transmitting in the UV region, highly absorbing in the visible region
8512	IR-absorbing sealing glass for Fe–Ni, lead-free (reed switches)
8516	IR-absorbing sealing glass for NiFe, lead-free, slow-evaporating (reed switches)
8531	Soft glass, Na-free, high lead-content, for low temperature encapsulation of semiconductor components (diodes)
8532	Soft glass, Na-free, high lead-content, for low-temperature encapsulation of semiconductor components (diodes)
8533	IR-absorbing sealing glass for Ni–Fe, lead- and potassium-free, slow-evaporating (reed switches)
8625	IR-absorbing biocompatible glass for (implantable) transponders
8650	Alkali-free sealing glass for molybdenum, especially for implosion diodes; high lead content
8651	Tungsten sealing glass for power diodes
8652	Tungsten sealing glass, low-melting, for power diodes
8656	Borofloat® 40, borosilicate float glass adapted for prestressing

[a] Also known as 8258, Estax®, low-potassium glass tubing for the manufacture of counting vials.

Table 3.4-11 Characteristic data of technical specialty glasses

Glass No.	Shapes produced[a]	Thermal expansion coefficient $\alpha_{(20/300)}$ (10^{-6}/K)	Transformation temperature T_g (°C)	Temperature at viscosity 10^{13} dPa s (°C)	$10^{7.6}$ dPa s (°C)	10^4 dPa s (°C)	Density at 25 °C (g/cm³)	Young's modulus (10^3 N/mm²)	Poisson's ratio μ	Heat conductivity λ at 90 °C (W/m K)
8095	TP	9.2	435	435	635	985	3.01	60	0.22	0.9
8245	MTRP	5.1	505	515	720	1040	2.31	68	0.22	1.2
8248	BP	3.1	445	490	740	1260	2.12	44	0.22	1.0
8250	MTBPC	5.0	490	500	720	1055	2.28	64	0.21	1.2
8252	TP	4.6	725	725	935	1250	2.63	81	0.24	1.1
8253	TP	4.7	785	790	1000	1315	2.65	83	0.23	1.1
8261	SP	3.7	720	725	950	1255	2.57	79	0.24	1.1
8326	MTP	6.6	560	565	770	1135	2.46	75	0.20	1.2
8330	MSTRPC	3.3	525	560	820	1260	2.23	63	0.20	1.12
8337B	TP	4.1	430	465	715	1090	2.21	51	0.22	1.0
8350	TRP	9.1	525	530	715	1040	2.50	73	0.22	1.1
8405	MTP	9.8	460	450	660	1000	2.51	65	0.21	1.0
8409	MTRP	4.1	745	740	950	1230	2.57	85	0.24	1.2
8412	TP	4.9	565	565	780	1165	2.34	73	0.20	1.2
8414	TP	5.4	560	560	770	1155	2.42	71	0.19	1.2
8415	TP	7.8	535	530	720	1050	2.50	74	0.21	1.1
8421	P	9.7	525	535	705	1000	2.59	74	0.22	1.0
8422	P	8.7	540	535	715	1010	2.46	76	0.21	1.1
8436	TRP	6.7	630	630	830	1110	2.76	85	0.22	1.1
8486	MP	4.1	555	580	820	1220	2.32	66	0.20	1.1
8487	TRP	3.9	525	560	775	1135	2.25	66	0.20	1.2
8488	M	4.3	545	560	800	1250	2.30	67	0.20	1.2
8490	MP	9.6	475	480	660	1000	2.61	70	0.22	1.0
8512	TP	9.0	445	460	665	980	2.53	68	0.22	1.0
8516	TP	8.9	440	445	650	990	2.56	72	0.21	1.1
8531	TP	9.0	440	430	590	830	4.34	52	0.24	0.57
8532	TP	8.8	430	425	565	760	4.47	56	0.24	0.7
8533	TP	8.7	475	480	645	915	2.57	79	0.21	1.1
8625	TP	9.0	510	520	710	1030	2.53	73	0.22	1.1
8650	TP	5.2	475	475	620	880	3.57	62	0.23	0.5
8651	TP	4.5	540	540	735	1040	2.87	59	0.24	0.9
8652	TP	4.5	495	490	640	915	3.18	58	0.25	0.9
8656	SP	4.1	590	600	850	1270	2.35	–	–	–

[a] Shapes produced: B = block glass; C = capillaries; M = molded glass (blown or pressed); P = powder, spray granulates, or sintered parts; R = rods; S = sheet glass; T = tubing

Table 3.4-11 Characteristic data of technical specialty glasses, cont.

T_{k100} (°C)	Logarithm of electrical volume resistivity (Ω cm) at 250 °C	350 °C	Dielectric properties at 1 MHz and 25 °C ε_r	tan δ (10^{-4})	Refractive index n_d ($\lambda_d = 587.6$ nm)	Stress-optical coefficient K (10^{-6} mm^2/N)	Classes of chemical stability against Water	Acid	Alkaline solution
330	9.6	7.6	6.6	11	1.556	3.1	3	2	3
215	7.4	5.9	5.7	80	1.488	3.8	3	4	3
–	12	10	4.3	10	1.466	5.2	3	3	3
375	10	8.3	4.9	22	1.487	3.6	3	4	3
660	–	12	6.1	11	1.538	3.3	1	3	2
630	–	11	6.6	15	1.547	2.7	1	2	2
585	–	–	5.8	14	1.534	3.1	1	4	2
210	7.3	6.0	6.4	65	1.506	2.8	1	1	2
250	8.0	6.5	4.6	37	1.473	4.0	1	1	2
315	9.2	7.5	4.7	22	1.476	4.1	3	4	3
200	7.1	5.7	7.2	70	1.514	2.7	3	1	2
280	8.5	6.9	6.5	45	1.505	2.8	5	3	2
530	12	10	6.1	23	1.543	2.9	1	4	3
215	7.4	6.0	5.7	80	1.492	3.4	1	1	2
200	7.1	5.6	6.3	107	1.523	2.2	1	2	2
180	6.7	5.3	7.1	113	1.521	3.2	2	2	2
255	8.1	6.4	7.4	43	1.526	2.7	3	3	2
205	7.3	5.8	7.3	60	1.509	2.9	2	3	3
245	7.9	6.5	7.9	75	1.564	2.9	1–2	1–2	1
230	7.5	6.1	5.1	40	1.487	3.8	1	1	2
300	8.3	6.9	4.9	36	1.479	3.6	4	3	3
200	7.1	5.8	5.4	96	1.484	3.2	1	1	2
235	7.7	6.1	6.7	32	1.52	–	3	2	2
320	9.5	7.5	6.5	21	1.510	3.0	3	1–2	2
250	8.1	6.4	6.5	25	1.516	3.0	3	1	2
450	11	9.8	9.5	9	1.700	2.2	1	4	3
440	11	9.4	10.2	9	1.724	1.7	1	4	3
200	7.0	5.5	6.9	55	1.527	3.0	1	2	2
210	7.2	5.8	7.1	68	1.525	–	3	1	2
–	–	–	7.6	33	1.618	2.8	1	4	3
–	11.2	10.0	6.0	31	1.552	3.6	1	4	3
–	–	–	6.9	35	1.589	3.4	1	4	3
265	8.3	6.8	5.5	51	1.493	3.6	1	1	1

3.4.5 Optical Glasses

Historically, optical glasses were developed to optimize imaging refractive optics in the visible part of the spectrum. With new mathematical tools, new light sources, and new detectors, the resulting lens systems, which used a combination of different glasses (or even sometimes crystalline materials), changed the requirement from high versatility in terms of material properties to homogeneity, precision, and reproducibility, and an extended spectral range in the ultraviolet and infrared regions; environmental aspects also became important. Consequently, the high number of glass types has been reduced.

3.4.5.1 Optical Properties

Refractive Index, Abbe Value, Dispersion, and Glass Designation

The most common identifying features used for characterizing an optical glass are the refractive index n_d in the middle range of the visible spectrum, and the Abbe value $v_d = (n_d - 1)/(n_F - n_C)$ as a measure of the dispersion. The difference $n_F - n_C$ is called the principal dispersion. The symbols have subscripts that identify spectral lines that are generally used to determine refractive indices; these spectral lines are listed in Table 3.4-12.

The quantities n_e and $v_e = (n_e - 1)/(n_{F'} - n_{C'})$ based on the e-line are usually used for specifying optical components.

For the comparison of glass types from different manufacturers, an abbreviated glass code is defined in the following way: the first three digits correspond to $n_d - 1$, the second three digits represent the Abbe value v_d, and, after a dot, three more digits characterize the density (see Table 3.4-13).

Glasses can be grouped into families in an n_d/v_d Abbe diagram (Fig. 3.4-25). These glass families differ in chemical composition as shown in Fig. 3.4-26.

Table 3.4-12 Wavelengths of a selection for frequently used spectral lines

Wavelength (nm)	Designation	Spectral line used	Element
2325.42		Infrared mercury line	Hg
1970.09		Infrared mercury line	Hg
1529.582		Infrared mercury line	Hg
1060.0		Neodymium glass laser	Nd
1013.98	t	Infrared mercury line	Hg
852.11	s	Infrared cesium line	Cs
706.5188	r	Red helium line	He
656.2725	C	Red hydrogen line	H
643.8469	C'	Red cadmium line	Cd
632.8		Helium–neon gas laser	He–Ne
589.2938	D	Yellow sodium line (center of the double line)	Na D
587.5618	d	Yellow helium line	He
546.0740	e	Green mercury line	Hg
486.1327	F	Blue hydrogen line	H
479.9914	F'	Blue cadmium line	Cd
435.8343	g	Blue mercury line	Hg
404.6561	h	Violet mercury line	Hg
365.0146	i	Ultraviolet mercury line	Hg
334.1478		Ultraviolet mercury line	Hg
312.5663		Ultraviolet mercury line	Hg
296.7278		Ultraviolet mercury line	Hg
280.43		Ultraviolet mercury line	Hg
248.00		Excimer laser	KrF
248.35		Ultraviolet mercury line	Hg
194.23			
193.00		Excimer laser	ArF

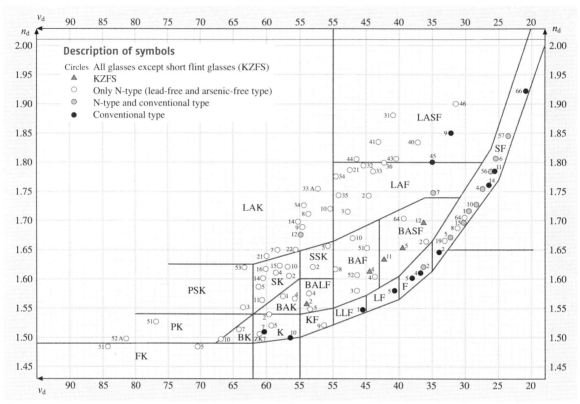

Fig. 3.4-25 Abbe diagram of glass families

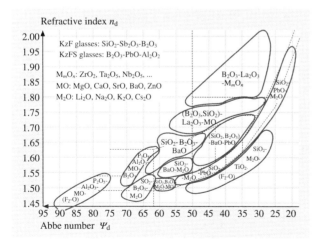

Fig. 3.4-26 Abbe diagram showing the chemical composition of the glass families

Table 3.4-13 Examples of glass codes

Glass type	n_d	v_d	Density ϱ (g cm^{-3})	Glass code	Remarks
N-SF6	1.80518	25.36	3.37	805 254.337	Lead- and arsenic-free glass
SF6	1.80518	25.43	5.18	805 254.518	Classical lead silicate glass

The designation of each glass type here is composed of an abbreviated family designation and a number. The glass families are arranged by decreasing Abbe value in the data tables (Table 3.4-14).

Table 3.4-14 gives an overview of the preferred optical glasses from Schott, Hoya, and Ohara. The glass types are listed in order of increasing refractive index n_d.

Table 3.4-14 Comparison of the preferred optical glasses from different manufacturers

Schott Code	Schott Glass type	Hoya Code	Hoya Glass type	Ohara Code	Ohara Glass type	Schott Code	Schott Glass type	Hoya Code	Hoya Glass type	Ohara Code	Ohara Glass type
434950	N-FK56									575415	S-TIL27
				439950	S-FPL53	580537	N-BALF4				
				456903	S-FPL52	581409	N-LF5	581407	E-FL5	581407	S-TIL25
487704	N-FK5	487704	FC5	487702	S-FSL5	581409	LF5	581409	FL5		
487845	N-FK51					583465	N-BAF3			583464	BAM3
497816	N-PK52	497816	FCD1	497816	S-FPL51			583594	BaCD12	583594	S-BAL42
498670	N-BK10					589613	N-SK5	589613	BaCD5	589612	S-BAL35
501564	K10					592683	N-PSK57				
508612	N-ZK7									593353	S-FTM16
511604	K7							594355	FF5		
		517522	CF6					596392	E-F8	596392	S-TIM8
		517524	E-CF6	517524	S-NSL36			596392	F8		
517642	N-BK7	517642	BSC7	516641	S-BSL7			603380	E-F5	603380	S-TIM5
				517696	S-APL1	603380	F5			603380	F5
		518590	E-C3	518590	S-NSL3	603606	N-SK14	603606	BaCD14	603607	S-BSM14
				521526	SSL5					603655	S-PHM53
522595	N-K5					606439	N-BAF4			606437	S-BAM4
				522598	S-NSL5	607567	N-SK2	607568	BaCD2	607568	S-BSM2
523515	N-KF9					609464	N-BAF52				
				529517	SSL2			613370	F3	613370	PBM3
529770	N-PK51					613443	KZFSN4	613443	ADF10	613443	BPM51
532488	N-LLF6	532488	FEL6			613445	N-KZFS4				
		532489	E-FEL6	532489	S-TIL6	613586	N-SK4	613587	BaCD4	613587	S-BSM4
540597	N-BAK2			540595	S-BAL12					614550	BSM9
		541472	E-FEL2	541472	S-TIL2	617366	F4				
		541472	FEL2							617628	S-PHM51
547536	N-BALF5					618498	N-SSK8			618498	S-BSM28
548458	LLF1	548458	FEL1					618634	PCD4	618634	S-PHM52
548458	N-LLF1	548458	E-FEL1	548458	S-TIL1	620364	N-F2	620363	E-F2	620363	S-TIM2
		551496	SbF1			620364	F2	620364	F2		
552635	N-PSK3					620603	N-SK16	620603	BaCD16	620603	S-BSM16
558542	N-KZFS2							620622	ADC1		
				560612	S-BAL50	620635	N-PSK53				
564608	N-SK11	564607	EBaCD11	564607	S-BAL41					621359	TIM11
		567428	FL6	567428	PBL26	621603	SK51				
569561	N-BAK4	569563	BaC4	569563	S-BAL14	622532	N-SSK2			622532	BSM22
569713	N-PSK58										
				571508	S-BAL2	623569	N-SK10	623570	EBaCD10	623570	S-BSM10
				571530	S-BAL3	623580	N-SK15	623582	BaCD15	623582	S-BSM15
573576	N-BAK1			573578	S-BAL11			624470	BaF8		

Table 3.4-14 Comparison of the preferred optical glasses from different manufacturers, cont.

Schott Code	Glass type	Hoya Code	Glass type	Ohara Code	Glass type	Schott Code	Glass type	Hoya Code	Glass type	Ohara Code	Glass type
		626357	F1			699301	N-SF15	699301	E-FD15	699301	S-TIM35
		626357	E-F1	626357	S-TIM1	699301	SF15	699301	FD15		
639421	N-KZFS11									700481	S-LAM51
				639449	S-BAM12			702412	BaFD7	702412	S-BAH27
639554	N-SK18	639554	BaCD18	639554	S-BSM18	704394	NBASF64				
		640345	E-FD7	640345	S-TIM27	706303	N-SF64				
640601	N-LAK21	640601	LaCL60	640601	S-BSM81	713538	N-LAK8	713539	LaC8	713539	S-LAL8
				641569	S-BSM93	717295	N-SF1	717295	E-FD1		
				643584	S-BSM36	717295	SF1	717295	FD1	717295	PBH1
648339	SF2	648339	FD2			717480	N-LAF3	717480	LaF3	717479	S-LAM3
		648338	E-FD2	648338	S-TIM22					720347	BPH8
		649530	E-BaCED20	649530	S-BSM71					720420	LAM58
651559	N-LAK22									720437	S-LAM52
		651562	LaCL2	651562	S-LAL54					720460	LAM61
652449	N-BAF51					720506	N-LAK10	720504	LaC10	720502	S-LAL10
652585	N-LAK7	652585	LaC7	652585	S-LAL7					722292	S-TIH18
654396	KZFSN5	654396	ADF50	654397	BPH5	724381	NBASF51			723380	S-BAH28
658509	N-SSK5	658509	BaCED5	658509	S-BSM25	724381	BASF51	724381	BaFD8		
				658573	S-LAL11					726536	S-LAL60
664360	N-BASF2					728284	SF10	728284	FD10		
				667330	S-TIM39	728285	N-SF10	728285	E-FD10	728285	S-TIH10
		667484	BaF11	667483	S-BAH11	729547	N-LAK34	729547	TaC8	729547	S-LAL18
				670393	BAH32			734515	TaC4	734515	S-LAL59
670471	N-BAF10	670473	BaF10	670473	S-BAH10					740283	PBH3W
								741276	FD13	740283	PBH3
				670573	S-LAL52			741278	E-FD13	741278	S-TIH13
673322	N-SF5	673321	E-FD5	673321	S-TIM25			741527	TaC2	741527	S-LAL61
673322	SF5	673322	FD5			743492	N-LAF35	743493	NbF1	743493	S-LAM60
		678507	LaCL9	678507	S-LAL56	744447	N-LAF2	744447	LaF2	744448	S-LAM2
678549	LAKL12					750350	LaFN7	750353	LaF7	750353	LAM7
678552	N-LAK12	678553	LaC12	678553	S-LAL12	750350	N-LAF7				
689312	N-SF8	689311	E-FD8	689311	S-TIM28	754524	N-LAK33	755523	TaC6	755523	S-YGH51
		689312	FD8			755276	N-SF4	755275	E-FD4	755275	S-TIH4
691547	N-LAK9	691548	LaC9	691548	S-LAL9	755276	SF4	755276	FD4		
		694508	LaCL5	694508	LAL58					756251	TPH55
694533	LAKN13	694532	LaC13	694532	S-LAL13			757478	NbF2	757478	S-LAM54
				695422	S-BAH54	762265	N-SF14	762265	FD140	762265	S-TIH14
		697485	LaFL2	697485	LAM59	762265	SF14	762266	FD14		
697554	N-LAK14	697555	LaC14	697555	S-LAL14					762401	S-LAM55
				697565	S-LAL64	772496	N-LAF34	772496	TaF1	772496	S-LAH66

Table 3.4-14 Comparison of preferred optical glasses, cont.

Schott		Hoya		Ohara	
Code	Glass type	Code	Glass type	Code	Glass type
785258	SF11	785258	FD11		
		785258	FD110	785257	S-TIH11
785261	SF56A				
785261	N-SF56	785261	FDS30	785263	S-TIH23
786441	N-LAF33	786439	NBFD11	786442	S-LAH51
				787500	S-YGH52
788475	N-LAF21	788475	TAF4	788474	S-LAH64
794454	N-LAF32			795453	S-LAH67
800423	N-LAF36	800423	NBFD12	800422	S-LAH52
801350	N-LASF45			801350	S-LAM66
				804396	S-LAH63
804466	N-LASF44	804465	TAF3	804466	S-LAH65
805254	N-SF6	805254	FD60	805254	S-TIH6
805254	SF6	805254	FD6		
		805396	NBFD3		
		806333	NBFD15		
806407	N-LASF43	806407	NBFD13	806409	S-LAH53
				808228	S-NPH1
		816445	TAFD10	816444	S-LAH54
		816466	TAF5	816466	S-LAH59
834374	N-LASF40	834373	NBFD10	834372	S-LAH60
835430	N-LASF41	835430	TAFD5	835427	S-LAH55
847238	N-SF57	847238	FDS90	847238	S-TIH53
847236	SFL57				
847238	SF57	847238	FDS9	847238	TIH53
850322	LASFN9				
				874353	S-LAH75
881410	N-LASF31				
		883408	TAFD30	883408	S-LAH58
901315	N-LASF46			901315	LAH78
923209	SF66	923209	E-FDS1		
				923213	PBH71
				1003283	S-LAH79
1022291	N-LASF35				

Formulas for Optical Characterization

The characterization of optical glasses through the refractive index and Abbe value alone is insufficient for high-quality optical systems. A more accurate description of the properties of a glass can be achieved with the aid of the relative partial dispersion.

Relative Partial Dispersion. The relative partial dispersion $P_{x,y}$ for the wavelengths x and y based on the blue F hydrogen line and red C hydrogen line is given by

$$P_{x,y} = (n_x - n_y)/(n_F - n_C) \,. \tag{4.15}$$

The corresponding value based on the blue F′ cadmium line and red C′ cadmium line is given by

$$P'_{x,y} = (n_x - n_y)/(n_{F'} - n_{C'}) \,. \tag{4.16}$$

Relationship Between the Abbe Value and the Relative Partial Dispersion. A linear relationship exists between the Abbe value and the relative partial dispersion for what are known as "normal glasses":

$$P_{x,y} \approx a_{xy} + b_{xy}\nu_d \,. \tag{4.17}$$

Deviation from the "Normal Line". All other glasses deviate from the "normal line" defined by $\Delta P_{x,y}$. For the selected wavelength pairs the ΔP-value are calculated from the following equations:

$$P_{x,y} = a_{xy} + b_{xy}\nu_d + \Delta P_{x,y} \,, \tag{4.18}$$

$$\Delta P_{C,t} = (n_C - n_t)/(n_F - n_C) \\ - (0.5450 + 0.004743\nu_d) \,, \tag{4.19}$$

$$\Delta P_{C,s} = (n_C - n_s)/(n_F - n_C) \\ - (0.4029 + 0.002331\nu_d) \,, \tag{4.20}$$

$$\Delta P_{F,e} = (n_F - n_e)/(n_F - n_C) \\ - (0.4884 - 0.000526\nu_d) \,, \tag{4.21}$$

$$\Delta P_{g,F} = (n_g - n_F)/(n_F - n_C) \\ - (0.6438 - 0.001682\nu_d) \,, \tag{4.22}$$

$$\Delta P_{i,g} = (n_i - n_g)/(n_F - n_C) \\ - (1.7241 - 0.008382\nu_d) \,. \tag{4.23}$$

The "normal line" has been determined based on value pairs of glasses types K7 and F2. The term $\Delta P_{x,y}$ quantitatively describes the deviation of the behavior of the dispersion from that of "normal glasses".

The Sellmeier dispersion formula for the refractive index,

$$n^2(\lambda) - 1 = B_1\lambda^2/(\lambda^2 - C_1) + B_2\lambda^2/(\lambda^2 - C_2) \\ + B_3\lambda^2/(\lambda^2 - C_3) \,, \tag{4.24}$$

can be derived from classical dispersion theory with the assumption of three resonance wavelengths. It is valid only for interpolation within a spectral region in which the refractive index has been measured. The vacuum wavelength λ in μm has to be used. The precision of the calculation achievable is generally better than 1×10^{-5}.

Temperature Dependence of the Refractive Index. The refractive index is also dependent on temperature. This temperature dependence is represented by $\Delta n_{\text{rel}}/\Delta T$ for an air pressure of 1013.3 hPa and $\Delta n_{\text{abs}}/\Delta T$ in vacuum. The following equation is derived from the Sellmeier formula and is valid with the given coefficients in the temperature range $-40\,°\text{C} < T < +80\,°\text{C}$ and within the wavelength range $435.8\,\text{nm} < \lambda < 643.8\,\text{nm}$:

$$\frac{dn_{\text{abs}}(\lambda, T)}{dT} = \frac{n^2(\lambda, T_0) - 1}{2n(\lambda, T_0)}$$
$$\times \left(D_0 + 2D_1 \Delta T + 3D_2 \Delta T^2 + \frac{E_0 + 2E_1 \Delta T}{\lambda^2 - \lambda_{\text{TK}}^2} \right), \quad (4.25)$$

where ΔT is the temperature difference in °C from 20 °C, and λ_{TK} is an effective resonance wavelength.

The changes in the refractive index and Abbe value caused by a change in the annealing rate are given by

$$n_d(h_x) = n_d(h_0) + m_{nd} \log(h_x/h_0), \quad (4.26)$$
$$v_d(h_x) = v_d(h_0) + m_{vd} \log(h_x/h_0), \quad (4.27)$$
$$m_{vd} = \frac{m_{nd} - v_d(h_0) m_{nF-nC}}{(n_F - n_C) + 2m_{nF-nC} \log(h_x/h_0)}, \quad (4.28)$$

where h_0 is the original annealing rate in °C/h, h_x is the new annealing rate in °C/h, m_{nd} is the annealing coefficient for the refractive index (Table 3.4-15), m_{vd} is the annealing coefficient for the Abbe value and m_{nF-nC} is the annealing coefficient for the principal dispersion. The last three quantities depend on the glass type.

The measurement accuracy of the Abbe value can be calculated using

$$\sigma(v_d) \approx \sigma(n_F - n_C) v_d/(n_F - n_C). \quad (4.29)$$

The accuracy of precision measurements of the refractive indices is better than $\pm 1 \times 10^{-5}$, and the accuracy of the dispersion is $\pm 3 \times 10^{-6}$. In the infrared wavelength range above 2 μm, the corresponding accuracies are $\pm 2 \times 10^{-5}$ and $\pm 5 \times 10^{-6}$.

Table 3.4-15 Annealing coefficients for selected glass types

Glass type	m_{nd}	m_{nF-nC}	m_{vd}
N-BK7	−0.00087	−0.000005	−0.0682
N-FK51	−0.00054	−0.000002	−0.0644
SF6	−0.00058	+0.000035	−0.0464
N-SF6	−0.0025	−0.000212	+0.0904

Transmission. The transmittance of glasses is limited by electronic excitations and light scattering in the UV, by vibronic excitations in the IR, and by reflections and impurity absorptions within the transmission window (in the visible part of the spectrum): Fig. 3.4-27. The UV absorption edge is temperature dependent. An example is shown in Fig. 3.4-28.

Spectral Internal Transmittance. The spectral internal transmittance is given by

$$\tau_{i\lambda} = \Phi_{e\lambda}/\Phi_{i\lambda}, \quad (4.30)$$

where $\Phi_{i\lambda}$ is the incident light intensity and $\Phi_{e\lambda}$ is the intensity at the exit.

Spectral Transmission. The spectral transmission is given by

$$\tau_\lambda = \tau_{i\lambda} P_\lambda, \quad (4.31)$$

where P_λ is the reflection factor.

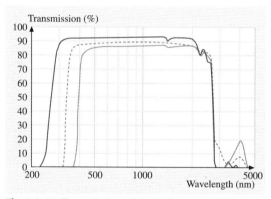

Fig. 3.4-27 Transmission of three glasses for a thickness of 5 mm: *Brown line* FK5; *dashed line* SF2; *gray line* SF11

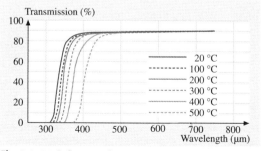

Fig. 3.4-28 Influence of temperature on the UV transmission of glass F2 for a thickness of 10 mm

Fresnel Reflectivity. For a light beam striking the surface perpendicularly, the Fresnel reflectivity is, independent of polarization,

$$R = (n-1)^2/(n+1)^2 \,. \tag{4.32}$$

Reflection Factor. The reflection factor, taking account of multiple reflection, is given by

$$P = (1-R)^2/(1-R^2) = 2n/(n^2+1) \,, \tag{4.33}$$

where n is the refractive index for the wavelength λ.

Conversion of Internal Transmittance to Another Layer Thickness. The conversion of data for internal transmittance to another sample thickness is accomplished by the use of the equation

$$\log \tau_{i1} / \log \tau_{i2} = d_1/d_2 \tag{4.34}$$

$$\text{or } \tau_{i2} = \tau_{i1}^{(d_2/d_1)} \,, \tag{4.35}$$

where τ_{i1} and τ_{i2} are the internal transmittances for the thicknesses d_1 and d_2, respectively.

Stress Birefringence. The change in optical path length for existing stress birefringence can be calculated from

$$\Delta s = (n_\parallel - n_\perp)d = (K_\parallel - K_\perp)d\sigma = K d\sigma \,, \tag{4.36}$$

where K is the stress optical coefficient, dependent on the glass type, d is the length of the light path in the sample, and σ is the mechanical stress (positive for tensile stress). If K is given in 10^{-6} mm^2/N, d is given in mm, and σ is measured in MPa = N/mm^2, Δs comes out in mm.

For the Pockels glass SF57, the stress optical coefficient K is close to 0 in the visible wavelength range.

Homogeneity. The homogeneity of the refractive index of a sample can be measured from the interferometrically measured wavefront deformation using the equation

$$\Delta n = \Delta W/2d$$
$$= \Delta W(\lambda) \times 633 \times 10^{-6}/(2d[\text{mm}]) \,, \tag{4.37}$$

where the wavefront deformation is in units of the wavelength and is measured using a test wavelength of 633 nm (He–Ne laser); ΔW is the wavefront deformation for double beam passage; and d is the thickness of the test piece. With special effort during melting and careful annealing, it is possible to produce pieces of glass having high homogeneity. The refractive-index homogeneity achievable for a given glass type depends on the volume and the form of the individual glass piece. Values of $\pm 5 \times 10^{-7}$ (class H5) cannot be achieved for all dimensions and glass types.

The properties of a selection of optical glasses are collected together in Table 3.4-16.

Internal Transmittance and Color Code

The internal transmittance, i.e. the light transmission excluding reflection losses, is closely related to the optical position of the glass type, according to general dispersion theory. This can be achieved, however, only by using purest raw materials and costly melting technology.

The internal transmittance of lead- and arsenic-free glasses, in which lead has been replaced by other elements, is markedly less than in the lead-containing predecessor glasses.

The limit of the transmission range of optical glasses towards the UV area is of special interest and is characterized by the position and slope of the UV absorption curve, which is described by a color code. The color code gives the wavelengths λ_{80} and λ_5, at which the transmission (including reflection losses) is 0.80 and 0.05, respectively, at 10 mm thickness. The color code 33/30 means, for example, $\lambda_{80} = 330$ nm and $\lambda_5 = 300$ nm.

3.4.5.2 Chemical Properties

The composition of optical glasses includes elements that reduce chemical resistance. For these glasses, five test methods are used to assess the chemical behavior of polished glass surfaces in typical applications. The test methods and classification numbers take the place of those described for technical glasses in Sect. 3.4.4. Data for optical properties are found in Table 3.4-16c.

Climatic Resistance (ISO/WD 13384): Division into Climatic Resistance Classes CR 1–4

Climatic resistance describes the behavior of optical glasses at high relative humidity and high temperatures. In the case of sensitive glasses, a cloudy film can appear that generally cannot be wiped off.

The classifications are based on the increase in transmission haze ΔH after a 30 h test period. The glasses in class CR 1 display no visible attack after being subjected to 30 h of climatic change.

Under normal humidity conditions, no surface attack should be expected during the fabrication and storage of optical glasses in class CR 1. On the other hand, the fabrication and storage of optical glasses in class CR 4 should be done with caution because these glasses are very sensitive to climatic influences.

Stain Resistance: Division into Stain Resistance Classes FR 0–5

The test procedure gives information about possible changes in the glass surface (stain formation) under the influence of lightly acidic water (for example perspiration and acidic condensates) without vaporization. Two test solutions are used. Test solution I is a standard acetate solution with pH = 4.6, for classes FR 0 to 3. Test solution II is a sodium acetate buffer solution with pH = 5.6, for classes FR 4 and FR 5.

Interference color stains develop as a result of decomposition of the surface of the glass by the test solution. The measure used for classifying the glasses is the time that elapses before the first brown–blue stain occurs at a temperature of 25 °C.

Stain resistance class FR 0 contains all glasses that exhibit virtually no interference colors even after 100 h of exposure to test solution I.

Glasses in classification FR 5 must be handled with particular care during processing.

Acid Resistance (ISO 8424: 1987): Division into Acid Resistance Classes SR 1–4, 5, and 51–53

Acid resistance classifies the behavior of optical glasses that come into contact with large quantities of acidic solutions (from a practical standpoint, these may be perspiration, laminating substances, carbonated water, etc.).

The time t required to dissolve a layer with a thickness of 0.1 μm serves as a measure of acid resistance. Two aggressive solutions are used in determining acid resistance. A strong acid (nitric acid, $c = 0.5$ mol/l, pH = 0.3) at 25 °C is used for the more resistant glass types. For glasses with less acid resistance, a weakly acidic solution with a pH value of 4.6 (standard acetate) is used, also at 25 °C.

Alkali Resistance (ISO 10629) and Phosphate Resistance (ISO 9689): Division into Alkali Resistance Classes AR 1–4 and Phosphate Resistance Classes PR 1–4

These two test methods indicate the resistance to aqueous alkaline solutions in excess and use the same classification scheme. The alkali resistance indicates the sensitivity of optical glasses when they are in contact with warm, alkaline liquids, such as cooling liquids used in grinding and polishing processes. The phosphate resistance describes the behavior of optical glasses during cleaning with phosphate-containing washing solutions (detergents).

The alkali resistance class AR is based on the time required to remove a layer of glass of thickness 0.1 μm in an alkaline solution (sodium hydroxide, $c = 0.01$ mol/l, pH = 12) at a temperature of 50 °C.

The phosphate resistance class PR is based on the time required to remove a layer of glass of thickness 0.1 mm in an alkaline phosphate-containing solution (pentasodium triphosphate, $Na_5P_3O_{10}$, $c = 0.01$ mol/l, pH = 10) at a temperature of 50 °C. The thickness is calculated from the weight loss per unit surface area and the density of the glass.

3.4.5.3 Mechanical Properties

Young's Modulus and Poisson's Ratio

The adiabatic Young's modulus E (in units of 10^3 N/mm^2) and Poisson's ratio μ have been determined at room temperature and at a frequency of 1 kHz using carefully annealed test samples. Data are listed in Table 3.4-16c. In most cases, the values decrease slightly with temperature.

The torsional modulus can be calculated from

$$G = E/[2(1+\mu)] \, . \tag{4.38}$$

The longitudinal sound velocity is

$$v_{\text{long}} = \sqrt{\frac{E(1-\mu)}{\varrho(1+\mu)(1-2\mu)}} \, , \tag{4.39}$$

where ϱ is the density.

Knoop Hardness

The Knoop hardness (HK) of a material is a measure of the residual surface changes after the application of pressure with a test diamond. The standard ISO 9385 describes the measurement procedure for glasses. In accordance with this standard, values for Knoop hardness HK are listed in the data sheets for a test force of 0.9807 N (corresponds to 0.1 kp) and an effective test period of 20 s. The test was performed on polished glass surfaces at room temperature. The data for hardness values are rounded to 10 HK 0.1/20. The microhardness is a function of the magnitude of the test force and decreases with increasing test force.

Table 3.4-16a Properties of optical glasses. Refractive index and Sellmeier constants

Glass type	Refractive index n_d	n_e	Abbe value v_d	v_e	Constants of the Sellmeier dispersion formula B_1	B_2	B_3	C_1	C_2	C_3
F2	1.62004	1.62408	36.37	36.11	$1.34533359 \times 10^{+00}$	$2.09073176 \times 10^{-01}$	$9.37357162 \times 10^{-01}$	$9.97743871 \times 10^{-03}$	$4.70450767 \times 10^{-02}$	$1.11886764 \times 10^{+02}$
K10	1.50137	1.50349	56.41	56.15	$1.15687082 \times 10^{+00}$	$6.42625444 \times 10^{-02}$	$8.72376139 \times 10^{-01}$	$8.09424251 \times 10^{-03}$	$3.86051284 \times 10^{-02}$	$1.04747730 \times 10^{+02}$
LASF35	2.02204	2.03035	29.06	28.84	$2.45505861 \times 10^{+00}$	$4.53006077 \times 10^{-01}$	$2.38513080 \times 10^{+00}$	$1.35670404 \times 10^{-02}$	$5.45803020 \times 10^{-02}$	$1.67904715 \times 10^{+02}$
LF5	1.58144	1.58482	40.85	40.57	$1.28035628 \times 10^{+00}$	$1.63505973 \times 10^{-01}$	$8.93930112 \times 10^{-01}$	$9.29854416 \times 10^{-03}$	$4.49135769 \times 10^{-02}$	$1.10493685 \times 10^{+02}$
LLF1	1.54814	1.55098	45.89	45.60	$1.23326922 \times 10^{+00}$	$1.16923839 \times 10^{-01}$	$8.62645379 \times 10^{-01}$	$8.85396812 \times 10^{-03}$	$4.36875155 \times 10^{-02}$	$1.04992168 \times 10^{+02}$
N-BAF10	1.67003	1.67341	47.11	46.83	$1.58514950 \times 10^{+00}$	$1.43559385 \times 10^{-01}$	$1.08521269 \times 10^{+00}$	$9.26681282 \times 10^{-03}$	$4.24489805 \times 10^{-02}$	$1.05613573 \times 10^{+02}$
N-BAF52	1.60863	1.61173	46.60	46.30	$1.43903433 \times 10^{+00}$	$9.67046052 \times 10^{-02}$	$1.09875818 \times 10^{+00}$	$9.07800128 \times 10^{-03}$	$5.08212080 \times 10^{-02}$	$1.05691856 \times 10^{+02}$
N-BAK4	1.56883	1.57125	55.98	55.70	$1.28834642 \times 10^{+00}$	$1.32817724 \times 10^{-01}$	$9.45395373 \times 10^{-01}$	$7.79980626 \times 10^{-03}$	$3.15631177 \times 10^{-02}$	$1.05965875 \times 10^{+02}$
N-BALF4	1.57956	1.58212	53.87	53.59	$1.31004128 \times 10^{+00}$	$1.42038259 \times 10^{-01}$	$9.64929351 \times 10^{-01}$	$7.96596450 \times 10^{-03}$	$3.30672072 \times 10^{-02}$	$1.09197320 \times 10^{+02}$
N-BASF64	1.70400	1.70824	39.38	39.12	$1.65554268 \times 10^{+00}$	$1.71319770 \times 10^{-01}$	$1.33664448 \times 10^{+00}$	$1.04485644 \times 10^{-02}$	$4.99394756 \times 10^{-02}$	$1.18961472 \times 10^{+02}$
N-BK7	1.51680	1.51872	64.17	63.96	$1.03961212 \times 10^{+00}$	$2.31792344 \times 10^{-01}$	$1.01046945 \times 10^{+00}$	$6.00069867 \times 10^{-03}$	$2.00179144 \times 10^{-02}$	$1.03560653 \times 10^{+02}$
N-FK56	1.43425	1.43534	94.95	94.53	$9.11957171 \times 10^{-01}$	$1.28580417 \times 10^{-01}$	$9.83146162 \times 10^{-01}$	$4.50933489 \times 10^{-03}$	$1.53515963 \times 10^{-02}$	$2.23961126 \times 10^{+02}$
N-KF9	1.52346	1.52588	51.54	51.26	$1.19286778 \times 10^{+00}$	$8.93346571 \times 10^{-02}$	$9.20819805 \times 10^{-01}$	$8.39154696 \times 10^{-03}$	$4.04010786 \times 10^{-02}$	$1.12572446 \times 10^{+02}$
N-KZFS2	1.55836	1.56082	54.01	53.83	$1.23697554 \times 10^{+00}$	$1.53569376 \times 10^{-01}$	$9.03976272 \times 10^{-01}$	$7.47170505 \times 10^{-03}$	$3.08053556 \times 10^{-02}$	$7.01731084 \times 10^{+01}$
N-LAF2	1.74397	1.74791	44.85	44.57	$1.80984227 \times 10^{+00}$	$1.57295550 \times 10^{-01}$	$1.09300370 \times 10^{+00}$	$1.01711622 \times 10^{-02}$	$4.42431765 \times 10^{-02}$	$1.00687748 \times 10^{+02}$
N-LAK33	1.75398	1.75740	52.43	52.20	$1.45796869 \times 10^{+00}$	$5.55403936 \times 10^{-01}$	$1.19938794 \times 10^{+00}$	$6.80545280 \times 10^{-03}$	$2.25253283 \times 10^{-02}$	$8.27543327 \times 10^{+01}$
N-LASF31	1.88067	1.88577	41.01	40.76	$1.71317198 \times 10^{+00}$	$7.18575109 \times 10^{-01}$	$1.72332470 \times 10^{+00}$	$8.19172228 \times 10^{-03}$	$2.97801704 \times 10^{-02}$	$1.38461313 \times 10^{+02}$
N-PK51	1.52855	1.53019	76.98	76.58	$1.15610775 \times 10^{+00}$	$1.53229344 \times 10^{-01}$	$7.85618966 \times 10^{-01}$	$5.85597402 \times 10^{-03}$	$1.94072416 \times 10^{-02}$	$1.40537046 \times 10^{+02}$
N-PSK57	1.59240	1.59447	68.40	68.01	$9.88511414 \times 10^{-01}$	$5.10855261 \times 10^{-01}$	$7.58837122 \times 10^{-01}$	$4.78397680 \times 10^{-03}$	$1.58020289 \times 10^{-02}$	$1.29709222 \times 10^{+02}$
N-SF1	1.71736	1.72308	29.62	29.39	$1.60865158 \times 10^{+00}$	$2.37725916 \times 10^{-01}$	$1.51530653 \times 10^{+00}$	$1.19654879 \times 10^{-02}$	$5.90589722 \times 10^{-02}$	$1.35521676 \times 10^{+02}$
N-SF56	1.78470	1.79179	26.10	25.89	$1.73562085 \times 10^{+00}$	$3.17487012 \times 10^{-01}$	$1.95398203 \times 10^{+00}$	$1.29624742 \times 10^{-02}$	$6.12884288 \times 10^{-02}$	$1.61559441 \times 10^{+02}$
N-SK16	1.62041	1.62286	60.32	60.08	$1.34317774 \times 10^{+00}$	$2.41144399 \times 10^{-01}$	$9.94317969 \times 10^{-01}$	$7.04687339 \times 10^{-03}$	$2.29005000 \times 10^{-02}$	$9.27508526 \times 10^{+01}$
N-SSK2	1.62229	1.62508	53.27	52.99	$1.43060270 \times 10^{+00}$	$1.53150554 \times 10^{-01}$	$1.01390904 \times 10^{+00}$	$8.23982975 \times 10^{-03}$	$3.33736841 \times 10^{-02}$	$1.06870822 \times 10^{+02}$
SF1	1.71736	1.72310	29.51	29.29	$1.55912923 \times 10^{+00}$	$2.84246288 \times 10^{-01}$	$9.68842926 \times 10^{-01}$	$1.21481001 \times 10^{-02}$	$5.34549042 \times 10^{-02}$	$1.12174809 \times 10^{+02}$
SF11	1.78472	1.79190	25.76	25.55	$1.73848403 \times 10^{+00}$	$3.11168974 \times 10^{-01}$	$1.17490871 \times 10^{+00}$	$1.36068604 \times 10^{-02}$	$6.15960463 \times 10^{-02}$	$1.21922711 \times 10^{+02}$
SF2	1.64769	1.65222	33.85	33.60	$1.40301821 \times 10^{+00}$	$2.31767504 \times 10^{-01}$	$9.39056586 \times 10^{-01}$	$1.05795466 \times 10^{-02}$	$4.93226978 \times 10^{-02}$	$1.12405955 \times 10^{+02}$
SF66	1.92286	1.93325	20.88	20.73	$2.07842233 \times 10^{+00}$	$4.07120032 \times 10^{-01}$	$1.76711292 \times 10^{+00}$	$1.80875134 \times 10^{-02}$	$6.79493572 \times 10^{-02}$	$2.15266127 \times 10^{+02}$
SK51	1.62090	1.62335	60.31	60.02	$1.44112715 \times 10^{+00}$	$1.43968387 \times 10^{-01}$	$8.81989862 \times 10^{-01}$	$7.58546975 \times 10^{-03}$	$2.87396017 \times 10^{-02}$	$9.46838154 \times 10^{+01}$
K7	1.51112	1.51314	60.41	60.15	$1.12735550 \times 10^{+00}$	$1.24412303 \times 10^{-01}$	$8.27101707 \times 10^{-01}$	$7.20341707 \times 10^{-03}$	$2.69835916 \times 10^{-02}$	$1.00384588 \times 10^{+02}$
N-SF6	1.80518	1.81266	25.36	25.16	$1.77931763 \times 10^{+00}$	$3.38149866 \times 10^{-01}$	$2.08734474 \times 10^{+00}$	$1.33714182 \times 10^{-02}$	$6.17533621 \times 10^{-02}$	$1.74017590 \times 10^{+02}$
SF6	1.80518	1.81265	25.43	25.24	$1.72448482 \times 10^{+00}$	$3.90104889 \times 10^{-01}$	$1.04572858 \times 10^{+00}$	$1.34871947 \times 10^{-02}$	$5.69318095 \times 10^{-02}$	$1.18557185 \times 10^{+02}$
N-FK51	1.48656	1.48794	84.47	84.07	$9.71247817 \times 10^{-01}$	$2.16901417 \times 10^{-01}$	$9.04651666 \times 10^{-01}$	$4.72301995 \times 10^{-03}$	$1.53575612 \times 10^{-02}$	$1.68681330 \times 10^{+02}$
Lithosil™ Q	1.45843	1.46004	67.87	67.67	$6.69422575 \times 10^{-01}$	$4.34583937 \times 10^{-01}$	$8.71694723 \times 10^{-01}$	$4.48011239 \times 10^{-03}$	$1.32847049 \times 10^{-02}$	$9.53414824 \times 10^{+01}$

Table 3.4-16b Data for dn/dT

Glass type	Data for dn/dT					
	$10^6 D_0$	$10^8 D_1$	$10^{11} D_2$	$10^7 E_0$	$10^{10} E_1$	λ_{TK} (µm)
F2	1.51	1.56	−2.78	9.34	10.4	0.250
K10	4.86	1.72	−3.02	3.82	4.53	0.260
LASF35	0.143	0.871	−2.71	10.2	15.0	0.263
LF5	−2.27	0.971	−2.83	8.36	9.95	0.228
LLF1	0.325	1.74	−6.12	6.53	2.58	0.233
N-BAF10	3.79	1.28	−1.42	5.84	7.60	0.220
N-BAF52	1.15	1.27	−0.508	5.64	6.38	0.238
N-BAK4	3.06	1.44	−2.23	5.46	6.05	0.189
N-BALF4	5.33	1.47	−1.58	5.75	6.58	0.195
N-BASF64	1.60	1.02	−2.68	7.87	9.65	0.229
N-BK7	1.86	1.31	−1.37	4.34	6.27	0.170
N-FK56	−20.4	−1.03	0.243	3.41	4.37	0.138
N-KF9	−1.66	0.844	−1.01	6.10	6.96	0.217
N-KZFS2	6.77	1.31	−1.23	3.84	5.51	0.196
N-LAF2	−3.64	0.920	−0.600	6.43	6.11	0.220
N-LAK33	2.57	1.16	−7.29	6.01	1.59	0.114
N-LASF31	2.29	0.893	−1.59	6.52	8.09	0.236
N-PK51	−19.8	−0.606	1.60	4.16	5.01	0.134
N-PSK57	−22.3	−0.560	0.997	4.47	5.63	–
N-SF1	−3.72	0.805	−1.71	8.98	13.4	0.276
N-SF56	−4.13	0.765	−1.12	9.90	15.7	0.287
N-SK16	−0.0237	1.32	−1.29	4.09	5.17	0.170
N-SSK2	5.21	1.34	−1.01	5.21	5.87	0.199
SF1	4.84	1.70	−4.52	13.8	12.6	0.259
SF11	11.2	1.81	−5.03	14.6	15.8	0.282
SF2	1.10	1.75	−1.29	10.8	10.3	0.249
SF66	–	–	–	–	–	–
SK51	−5.63	0.738	−6.20	3.91	2.64	0.230
K7	−1.67	0.880	−2.86	5.42	7.81	0.172
N-SF6	−4.93	0.702	−2.40	9.84	15.4	0.290
SF6	6.69	1.78	−3.36	17.7	17.0	0.272
N-FK51	−18.3	−0.789	−0.163	3.74	3.46	0.150
Lithosil™Q	20.6	2.51	−2.47	3.12	4.22	0.160

Table 3.4-16c Chemical and physical data

Glass type	Stress-optical coefficient K (10^{-6} mm²/N)	Chemical properties					Density (g/cm³)	Viscosity (dPa s) at temperature			Thermal properties		Thermal expansion		Mechanical properties		
		CR	FR	SR	AR	PR		$10^{14.5}$ (°C)	10^{13} (°C)	$10^{7.6}$ (°C)	Heat capacity c_p (J/gK)	Heat conductivity λ (W/mK)	$\alpha_{(30/70)}$ (10^{-6}/K)	$\alpha_{(20/300)}$ (10^{-6}/K)	Young's modulus E (10^3 N/mm²)	Poisson's ratio μ	Knoop hardness HK
F2	2.81	1	0	1	2.3	1.3	3.61	432	421	593	0.557	0.780	8.20	9.20	57	0.220	420
K10	3.12	1	0	1	1	1.2	2.52	459	453	691	0.770	1.120	6.50	7.40	65	0.190	470
LASF35	0.73	1	0	1.3	1	1.3	5.41	774	–	–	0.445	0.920	7.40	8.50	132	0.303	810
LF5	2.83	2	0	1	2.3	2	3.22	419	411	585	0.657	0.866	9.10	10.60	59	0.223	450
LLF1	3.05	1	0	1	2	1	2.94	448	426	628	0.650	–	8.10	9.20	60	0.208	450
N-BAF10	2.37	1	0	4.3	1.3	1	3.75	660	652	790	0.560	0.780	6.18	7.04	89	0.271	620
N-BAF52	2.42	1	0	1	1.3	1	3.05	594	596	723	0.680	0.960	6.86	7.83	86	0.237	600
N-BAK4	2.90	1	0	1.2	1	1	3.05	581	569	725	0.680	0.880	6.99	7.93	77	0.240	550
N-BALF4	3.01	1	0	1	1	1	3.11	578	584	661	0.690	0.850	6.52	7.41	77	0.245	540
N-BASF64	2.38	1	0	3.2	1.2	1	3.20	582	585	712	–	–	7.30	8.70	105	0.264	650
N-BK7	2.77	2	0	1	2	2.3	2.51	557	557	719	0.858	1.114	7.10	8.30	82	0.206	610
N-FK56	0.68	1	0	52.3	4.3	4.3	3.54	422	416	–	0.750	0.840	–	16.16	70	0.293	350
N-KF9	2.74	1	0	1	1	1	2.50	476	476	640	0.860	1.040	9.61	10.95	66	0.225	480
N-KZFS2	4.02	1	4	52.3	4.3	4.2	2.55	491	488	600	0.830	0.810	4.43	5.43	66	0.266	490
N-LAF2	1.42	2	3	52.2	1	2.2	4.30	653	645	742	0.510	0.670	8.06	9.10	94	0.288	530
N-LAK33	1.49	1	1	51.3	1	2.3	4.26	652	648	–	0.554	0.900	6.00	7.00	124	0.291	780
N-LASF31	1.10	1	0	2	1	1	5.41	758	756	–	–	0.910	6.80	7.70	124	0.299	770
N-PK51	0.54	2	0	51.2	3.3	4.3	3.96	496	486	–	–	–	12.70	14.40	74	0.295	400
N-PSK57	0.13	1	0	51.3	1.2	4.3	4.48	497	499	–	0.490	0.560	13.17	14.75	69	0.298	370
N-SF1	2.72	1	0	1	1	1	3.03	553	554	660	0.750	1.000	9.13	10.54	90	0.250	540
N-SF56	2.87	1	0	1	1.3	1	3.28	592	585	691	0.700	0.940	8.70	10.00	91	0.255	560
N-SK16	1.90	4	4	53.3	3.3	3.2	3.58	636	633	750	0.578	0.818	6.30	7.30	89	0.264	600
N-SSK2	2.51	1	0	1.2	1	1	3.53	653	655	801	0.580	0.810	5.81	6.65	82	0.261	570
SF1	1.80	2	1	3.2	2.3	3	4.46	417	415	566	0.431	0.737	8.10	8.80	56	0.232	390
SF11	1.33	1	0	1	1.2	1	4.74	503	500	635	0.498	0.735	6.10	6.80	66	0.235	450
SF2	2.62	1	0	2	2.3	2	3.86	441	428	600	0.340	0.530	8.40	9.20	55	0.227	410
SF66	–1.20	2	5	53.4	2.3	4.2	6.03	384	385	482	–	–	9.01	11.48	51	0.258	310
SK51	1.47	2	3	52.3	1.3	4.3	3.52	597	579	684	–	–	8.90	10.10	75	0.291	450
K7	2.95	3	0	2	1	2.3	2.53	513	–	712	0.69	0.96	8.4	9.7	69	0.214	520
N-SF6	2.82	1	0	2	2.3	3	3.37	594	591	694	0.389	0.673	9.03	10.39	93	0.262	550
SF6	0.65	2	3	51.3	2.3	3.3	5.18	423	410	538	0.636	0.911	8.1	9	55	0.244	370
N-FK51	0.70	2	0	52.3	2.2	4.3	3.73	420	403	–	–	–	13.3	15.3	81	0.293	430
Lithosil™ Q	3.40	1	–	1	1	–	2.20	980	1080	1600	0.790	1.310	0.50	–	72	0.170	580

Table 3.4-16d Internal transmission and color code

Glass type	Color code	Internal transmission measured for 25 mm sample thickness at wavelength λ (nm)												
		2500	2325	1970	1530	1060	700	660	620	580	546	500	460	436
F2	35/32	0.610	0.700	0.890	0.990	0.998	0.998	0.996	0.997	0.997	0.997	0.996	0.993	0.991
K10	33/30	0.520	0.630	0.850	0.983	0.996	0.997	0.994	0.993	0.993	0.992	0.991	0.990	0.988
LASF35	–/37	0.690	0.880	0.972	0.992	0.990	0.978	0.970	0.962	0.950	0.920	0.810	0.630	0.470
LF5	34/31	–	0.660	0.870	0.992	0.998	0.998	0.998	0.998	0.997	0.997	0.996	0.995	0.994
LLF1	33/31	0.500	0.610	0.840	0.990	0.996	0.997	0.996	0.996	0.997	0.997	0.996	0.996	0.996
N-BAF10	39/35	0.450	0.680	0.920	0.980	0.994	0.994	0.990	0.991	0.990	0.990	0.981	0.967	0.954
N-BAF52	39/35	0.390	0.630	0.890	0.975	0.994	0.993	0.990	0.989	0.990	0.989	0.980	0.967	0.954
N-BAK4	36/33	0.540	0.710	0.900	0.982	0.995	0.997	0.995	0.995	0.996	0.996	0.994	0.989	0.988
N-BALF4	37/33	0.580	0.740	0.920	0.984	0.993	0.997	0.995	0.995	0.996	0.995	0.993	0.986	0.983
N-BASF64	40/35	0.450	0.670	0.900	0.970	0.985	0.970	0.955	0.949	0.949	0.950	0.940	0.920	0.900
N-BK7	33/29	0.360	0.560	0.840	0.980	0.997	0.996	0.994	0.994	0.995	0.996	0.994	0.993	0.992
N-FK56	33/28	–	–	0.979	0.991	0.996	0.996	0.996	0.996	0.996	0.996	0.996	0.996	0.995
N-KF9	37/34	0.300	0.430	0.740	0.981	0.995	0.997	0.995	0.994	0.996	0.996	0.994	0.990	0.988
N-KZFS2	34/30	0.040	0.260	0.800	0.940	0.991	0.996	0.994	0.994	0.994	0.994	0.992	0.987	0.981
N-LAF2	40/34	0.400	0.690	0.930	0.990	0.997	0.996	0.993	0.992	0.993	0.994	0.983	0.962	0.940
N-LAK33	39/32	0.090	0.400	0.850	0.975	0.995	0.991	0.990	0.990	0.990	0.990	0.987	0.977	0.967
N-LASF31	45/32	0.540	0.810	0.960	0.992	0.993	0.994	0.994	0.993	0.993	0.990	0.973	0.940	0.910
N-PK51	35/29	0.890	0.920	0.965	0.985	0.992	0.991	0.991	0.992	0.994	0.995	0.993	0.989	0.987
N-PSK57	34/29	–	–	0.950	0.970	0.982	0.996	0.996	0.996	0.996	0.996	0.992	0.991	0.991
N-SF1	41/36	0.460	0.580	0.850	0.973	0.995	0.990	0.986	0.987	0.990	0.986	0.968	0.940	0.910
N-SF56	44/37	0.590	0.680	0.900	0.981	0.996	0.986	0.981	0.981	0.983	0.976	0.950	0.910	0.860
N-SK16	36/30	0.260	0.540	0.880	0.973	0.995	0.996	0.994	0.993	0.994	0.994	0.991	0.984	0.981
N-SSK2	37/33	0.500	0.720	0.930	0.981	0.992	0.996	0.994	0.993	0.995	0.995	0.992	0.985	0.980
SF1	39/34	0.650	0.730	0.900	0.985	0.996	0.996	0.995	0.995	0.996	0.996	0.993	0.984	0.976
SF11	44/39	0.610	0.700	0.930	0.982	0.997	0.993	0.991	0.991	0.991	0.989	0.976	0.940	0.860
SF2	37/33	0.620	0.710	0.880	0.985	0.996	0.996	0.994	0.995	0.995	0.995	0.993	0.988	0.982
SF66	48/38	0.700	0.740	0.920	0.990	0.995	0.990	0.989	0.989	0.988	0.985	0.965	0.890	0.770
SK51	36/31	0.270	0.520	0.830	0.959	0.993	0.993	0.993	0.993	0.993	0.993	0.990	0.981	0.975
K7	33/30	0.340	0.500	0.790	0.980	0.994	0.996	0.995	0.995	0.994	0.994	0.993	0.990	0.990
N-SF6	45/37	0.850	0.880	0.962	0.994	0.994	0.987	0.980	0.979	0.980	0.970	0.940	0.899	0.850
SF6	42/36	0.730	0.780	0.930	0.990	0.996	0.996	0.995	0.995	0.995	0.994	0.989	0.972	0.940
N-FK51	34/28	0.750	0.840	0.940	0.980	0.994	0.995	0.995	0.996	0.997	0.997	0.996	0.993	0.992
Lithosil™Q	17/16	0.780	–	–	–	–	–	–	–	–	–	–	–	–

Internal transmission measured for 25 mm sample thickness at wavelength λ (nm)														
420	405	400	390	380	370	365	350	334	320	310	300	290	248	200	193
0.990	0.986	0.984	0.977	0.963	0.940	0.920	0.780	0.210	–	–	–	–	–	–	–
0.988	0.987	0.986	0.982	0.973	0.966	0.958	0.910	0.720	0.310	0.130	0.020	–	–	–	–
0.320	0.170	0.120	0.050	0.010	–	–	–	–	–	–	–	–	–	–	–
0.993	0.992	0.992	0.984	0.973	0.961	0.954	0.880	0.570	0.040	–	–	–	–	–	–
0.995	0.994	0.993	0.992	0.988	0.984	0.981	0.955	0.810	0.300	0.010	–	–	–	–	–
0.940	0.900	0.880	0.800	0.660	0.440	0.310	0.010	–	–	–	–	–	–	–	–
0.938	0.900	0.880	0.800	0.650	0.370	0.210	–	–	–	–	–	–	–	–	–
0.987	0.983	0.980	0.967	0.940	0.890	0.840	0.550	0.070	–	–	–	–	–	–	–
0.981	0.970	0.964	0.940	0.900	0.820	0.750	0.380	–	–	–	–	–	–	–	–
0.880	0.840	0.820	0.750	0.610	0.370	0.220	–	–	–	–	–	–	–	–	–
0.993	0.993	0.992	0.989	0.983	0.977	0.971	0.920	0.780	0.520	0.250	0.050	–	–	–	–
0.994	0.996	0.996	0.995	0.992	0.985	0.975	0.920	0.760	0.460	0.210	0.060	0.010	–	–	–
0.985	0.975	0.965	0.940	0.880	0.770	0.680	0.210	–	–	–	–	–	–	–	–
0.975	0.967	0.963	0.950	0.930	0.910	0.890	0.800	0.590	0.240	0.030	–	–	–	–	–
0.915	0.865	0.840	0.760	0.630	0.430	0.310	0.025	–	–	–	–	–	–	–	–
0.954	0.928	0.910	0.860	0.790	0.690	0.630	0.400	0.140	0.020	–	–	–	–	–	–
0.880	0.840	0.820	0.750	0.650	0.530	0.460	0.210	0.040	0.020	–	–	–	–	–	–
0.986	0.985	0.984	0.977	0.965	0.940	0.910	0.750	0.430	0.120	0.030	–	–	–	–	–
0.991	0.991	0.992	0.992	0.989	0.975	0.965	0.880	0.680	0.380	0.130	0.020	–	–	–	–
0.870	0.760	0.700	0.520	0.250	0.030	–	–	–	–	–	–	–	–	–	–
0.780	0.640	0.570	0.370	0.130	–	–	–	–	–	–	–	–	–	–	–
0.979	0.974	0.970	0.956	0.930	0.890	0.860	0.700	0.400	0.110	0.020	–	–	–	–	–
0.975	0.963	0.954	0.920	0.860	0.750	0.670	0.250	–	–	–	–	–	–	–	–
0.961	0.930	0.920	0.870	0.790	0.640	0.500	0.030	–	–	–	–	–	–	–	–
0.700	0.340	0.200	0.010	–	–	–	–	–	–	–	–	–	–	–	–
0.975	0.962	0.954	0.920	0.870	0.790	0.720	0.370	–	–	–	–	–	–	–	–
0.610	0.340	0.240	0.050	–	–	–	–	–	–	–	–	–	–	–	–
0.971	0.963	0.958	0.940	0.910	0.850	0.800	0.600	0.300	0.100	0.030	–	–	–	–	–
0.990	0.990	0.990	0.988	0.983	0.976	0.971	0.940	0.780	0.420	0.100	–	–	–	–	–
0.780	0.640	0.570	0.370	0.140	–	–	–	–	–	–	–	–	–	–	–
0.900	0.810	0.760	0.620	0.370	0.100	0.020	–	–	–	–	–	–	–	–	–
0.992	0.993	0.993	0.992	0.988	0.976	0.963	0.875	0.630	0.300	0.120	0.035	0.010	–	–	–
–	–	–	–	–	–	–	–	–	–	–	–	–	0.995	0.990	0.980

Viscosity

As explained in the introduction, glasses pass through three viscosity ranges between the melting temperature and room temperature: the melting range, the super-cooled melt range, and the solidification range. The viscosity increases during the cooling of the melt, starting from $10^0 - 10^4$ dPa s. A transition from a liquid to a plastic state is observed between 10^4 and 10^{13} dPa s.

The softening point, i.e. the temperature where the viscosity is $10^{7.6}$ dPa s, identifies the plastic range in which glass parts rapidly deform under their own weight. The glass structure can be described as solidified or "frozen" above 10^{13} dPa s. At this viscosity, the internal stresses in glass anneal out equalize in approximately 15 min. The temperature at which the viscosity is 10^{13} dPa s is called the upper annealing point, and is important for the annealing of glasses.

In accordance with ISO 7884-8, the rate of change of the relative linear thermal expansion can be used to determine the transformation temperature T_g, which is close to the temperature at which the viscosity is 10^{13} dPa s.

Precision optical surfaces may deform and refractive indices may change if a temperature of $T_g - 200$ K is exceeded during any thermal treatment.

Coefficient of Linear Thermal Expansion

The typical curve of the linear thermal expansion of a glass begins with an increase in slope from absolute zero to approximately room temperature. Then a nearly linear increase to the beginning of the plastic behavior follows. The transformation range is distinguished by a distinct bending of the expansion curve, which results from the increasing structural rearrangement in the glass. Above this range, the expansion again exhibits a nearly linear increase, but with a noticeably greater slope.

Two averaged coefficients of linear thermal expansion α are usually given: $\alpha_{30/70}$, averaged from $-30\,°\mathrm{C}$ to $+70\,°\mathrm{C}$, which is the relevant value for room temperature; and $\alpha_{20/300}$, averaged from $+20\,°\mathrm{C}$ to $+300\,°\mathrm{C}$, which is the standard international value. These values are listed in Table 3.4-16.

3.4.5.4 Thermal Properties

Thermal Conductivity

The range of values for the thermal conductivity of glasses at room temperature extends from 1.38 W/m K (pure vitreous silica) to about 0.5 W/m K (high-lead-content glasses). The most commonly used silicate glasses have values between 0.9 and 1.2 W/m K. All data in Table 3.4-16c are given for a temperature of $90\,°\mathrm{C}$, with an accuracy of $\pm 5\%$.

Specific Thermal Capacity

The mean isobaric specific heat capacities c_p ($20\,°\mathrm{C}$; $100\,°\mathrm{C}$) listed in Table 3.4-16c were measured from the heat transfer from a hot glass sample at $100\,°\mathrm{C}$ into a liquid calorimeter at $20\,°\mathrm{C}$. The values of $c_p(20\,°\mathrm{C};\,100\,°\mathrm{C})$ and also of the true thermal capacity $c_p(20\,°\mathrm{C})$ for silicate glasses range from 0.42 to 0.84 J/g K.

3.4.6 Vitreous Silica

Vitreous silica has a unique set of properties. It is produced either from natural quartz by fusion or, if extreme purity is required, by chemical vapor deposition or via a sol–gel routes. Depending on the manufacturing process, variable quantities impurities are incorporated in the ppm or ppb range, such as Fe, Mg, Al, Mn, Ti, Ce, OH, Cl, and F. These impurities and radiation-induced defects, as well as complexes of impurities and defects, and also overtones, control the UV and IR transmittance. In the visible part of the spectrum, Rayleigh scattering from thermodynamically caused density fluctuations dominates. Defects are also responsible for the damage threshold under radiation load, and for fluorescence. The refractive index n and the absorption constant K as a function of wavelength are found in Fig. 3.4-29.

The highest transmittance is required for applications in optical communication networks, in optics for lithography, and in high-power laser physics. For certain applications, for example to increase the refractive index in the IR in fiber optics, the silica is "doped" with GeO_2, P_2O_5, B_2O_3, etc. in the range of $5-10\%$. In such cases the scattering loss increases owing to concentration fluctuations.

There are also many technical applications which make use of the chemical inertness, light weight, high temperature stability, thermal-shock resistance, and low thermal expansion of vitreous silica. A very low thermal

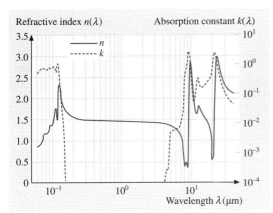

Fig. 3.4-29 Measured optical constants $n(\lambda)$ and $k(\lambda)$ of vitreous silica according to [4.16]

Table 3.4-17 Electrical properties of vitreous silica (Lithosil™)

Dielectric constant ε_r	3.8 ± 0.2
Dielectric loss angle φ	$89.92° \pm 0.03°$ at 25 °C and 1 MHz
$\tan \delta$ ($\delta = 90° - \varphi$)	$14 \pm 5 \times 10^{-4}$
Electrical resistivity	1.15×10^{18} (Ω cm) at 20 °C

expansion is obtained in ULE glass (Corning "ultralow expansion" glass) by doping with $\approx 9\%$ TiO$_2$.

3.4.6.1 Properties of Synthetic Silica

The precise data for materials from various suppliers differ slightly, depending on the thermal history and impurity concentration. The data listed in Table 3.4-16a–d and in Table 3.4-17, are for Lithosil™Q0 (Schott Lithotec). The various quantities are defined in the same way as for optical glasses, as described in Sect. 3.4.5.

3.4.6.2 Gas Solubility and Molecular Diffusion

The relatively open structure of vitreous silica provides space for the incorporation and diffusion of molecular species. The data in the literature are not very consistent; Table 3.4-18 should serve as an orientation.

The pressure dependence of the solubility is small up to about 100 atm.

The diffusion coefficient depends on temperature as

$$D = D_0 T \exp(-Q/RT) \,. \tag{4.40}$$

Water can react with the silica network:

$$H_2O + Si-O-Si = 2\,Si-OH \,. \tag{4.41}$$

The reaction has a strong influence on the concentration and apparent diffusion of dissolved molecular water.

Table 3.4-18 Solubility and diffusion of molecular gases in vitreous silica (Lithosil™)

Gas	Molecular diameter (nm)	c_{glass}/c_{gas} at 200–1000 °C	Dissolved molecules S (cm^{-3} atm^{-1}) at 200 °C	Diffusion coefficient D_0 (cm^2/s) 25 °C	1000 °C	Activation energy Q (kJ/mole)
Helium	0.20	0.025	3.9×10^{-17}	2.4×10^{-8}	5.5×10^{-5}	20
Neon	0.24	0.019	3.1×10^{-17}	5.0×10^{-12}	2.5×10^{-6}	37
Hydrogen	0.25	0.03	4.7×10^{-17}	2.2×10^{-11}	7.3×10^{-6}	36
Argon	0.32	0.01	1.5×10^{-17}	–	1.4×10^{-9}	111
Oxygen	0.32	0.01	1.5×10^{-17}	–	6.6×10^{-9}	105
Water	0.33	–	–	–	$\approx 3.0 \times 10^{-7}$	71
Nitrogen	0.34	–	–	–	–	110
Krypton	0.42	–	–	–	–	≈ 190
Xenon	0.49	–	–	–	–	≈ 300

3.4.7 Glass-Ceramics

Glass-ceramics are distinguished from glasses and from ceramics by the characteristics of their manufacturing processes (see introduction to this chapter 3.4) as well as by their physico-chemical properties.

They are manufactured in two principal production steps. In the first step, a batch of exactly defined composition is melted (as for a conventional glass). The composition is determined by the desired properties of the endproduct and by the necessary working properties of the glass. After melting, the product is shaped by pressing, blowing, rolling, or casting, and then annealed. In this second step, "glassy" articles are partly crystallized by use of a specific temperature–time program between 800 and 1200 °C (this program must be defined for each composition). Apart from the crystalline phase, with crystals 0.05–5 μm in size, this material contains a residual glass phase that amounts to 5–50% of the volume.

In the temperature range between 600 and 700 °C, small amounts of nucleating agents (e.g. TiO_2, ZrO_2, or F) induce precipitation of crystal nuclei. When the temperature is increased, crystals grow on these nuclei. Their type and properties, as well as their number and size, are predetermined by the glass composition and the annealing program. By selection of an appropriate program, either transparent, slightly opaque, or highly opaque, nontransparent glass-ceramics can be produced. Unlike conventional ceramics, these glass ceramics are fully dense and pore-free.

Like the composition of glasses, the composition of glass-ceramics is highly variable. Some well-known compositions lie within the following systems: $Li_2O-Al_2O_3-SiO_2$, $MgO-Al_2O_3-SiO_2$, and $CaO-P_2O_5-Al_2O_3-SiO_2$.

Glass-ceramics of the $Li_2O-Al_2O_3-SiO_2$ system, which contain small amounts of alkali and alkaline-earth oxides, as well as TiO_2 and ZrO_2 as nucleating agents, have achieved the greatest commercial importance. On the basis of this system, glass-ceramics with coefficients of linear thermal expansion near to zero can be produced (Fig. 3.4-30 and Table 3.4-19). This exceptional property results from the bonding of crystalline constituents (such as solid solutions of h-quartz, h-eucryptite, or h-spodumene) which have negative coefficients of thermal expansion with the residual glass phase of the system, which has a positive coefficient of thermal expansion.

Such "$\alpha = 0$ glass-ceramics" can be subjected to virtually any thermal shock or temperature variation below 700 °C. Wall thickness, wall thickness differences, and complicated shapes are of no significance.

Another technical advantage is the exceptionally high dimensional and shape stability of objects made from these materials, even when the objects are subjected to considerable temperature variations.

The Zerodur® glass-ceramic, whose coefficient of linear thermal expansion at room temperature can be kept at $\leq 0.05 \times 10^{-6}$/K (Table 3.4-19), was especially developed for the production of large mirror blanks for astronomical telescopes. Zerodur® has further applications in optomechanical precision components such as length standards, and mirror spacers in lasers. With

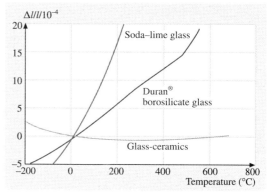

Fig. 3.4-30 Thermal expansion of glass-ceramics compared with borosilicate glass 3.3 and soda–lime glass

Table 3.4-19 Coefficient of linear thermal expansion α, density, and elastic properties of Zerodur® and Ceran® glass-ceramics

	Zerodur®	Ceran®	Units	Product class
$\alpha_{0/50}$	0 ± 0.05	–	10^{-6}/K	1
	0 ± 0.1	–	10^{-6}/K	2
	0 ± 0.15	–	10^{-6}/K	3
$\alpha_{20/300}$	+0.1	−0.2	10^{-6}/K	
$\alpha_{20/500}$	–	−0.01	10^{-6}/K	
$\alpha_{20/600}$	+0.2	–	10^{-6}/K	
$\alpha_{20/700}$	–	+0.15	10^{-6}/K	
Density	2.53	2.56	g/cm³	
Young's modulus E	91 × 10³	92 × 10³	N/mm²	
Poisson's ratio μ	0.24	0.24		

a length aging coefficient A (where $L = L_0(1 + A\Delta t)$, Δt = time span) below $1 \times 10^{-7}/y$, Zerodur® has excellent longitudinal stability.

The Ceran® glass-ceramic is colored and is designed for applications in cooker surface panels.

As in glasses, the variability of the composition can be used to design very different sets of properties of glass-ceramics. Some examples are:

- photosensitive, etchable glass-ceramics based on Ag doping: Foturan (Schott), and Fotoform and Fotoceram (Corning);
- machinable glass-ceramics based on mica crystals, for example for electronic packaging: Macor and Dicor (Corning), and Vitronit (Vitron and Jena);
- glass-ceramics used as substrates for magnetic disks, based on spinel or gahnite crystals, resulting in a very high elastic modulus and thus stiffness: Neoceram (NEG), and products from Corning and Ohara;
- glass-ceramics with extremely good weathering properties for architectural applications: Neoparies (NEG) and Cryston (Asahi Glass);
- biocompatible, bioactive glass-ceramics based on apatite and orthophosphate crystals for dental restoration or bone replacement in medicine: Cerabone (NEG), Bioverit (Vitron), Ceravital, IPS Empress, etc.;
- highly transparent glass-ceramics and glass-ceramics with specific dopings for temperature-resistant fiber optic components, high-temperature loaded color filters, and luminescent solar collectors.

An excellent overview and many details can be found in [4.4].

3.4.8 Glasses for Miscellaneous Applications

3.4.8.1 Sealing Glasses

Glasses are very well suited for the production of mechanically reliable, vacuum-tight fusion seals with metals, ceramics, and mica. Some particularly favorable properties are the viscosity behavior of glass and the direct wettability of many crystalline materials by glasses. As a result, the production technology for such seals is characterized by uncomplicated procedures with few, easily manageable, well-controllable process steps.

A necessary condition for the stability and mechanical strength of glass seals is the limitation of the mechanical stress in the glass component at temperatures encountered during production and use. To ensure "sealability" (which means that the thermal contractions of the two sealing components match each other below the transformation temperature of the glass), glasses of special compositions, called sealing glasses, have been developed. Apart from sealability, such glasses must very often fulfill other requirements such as high electrical insulation or special optical properties. The sealability can be tested and evaluated with sufficient accuracy and certainty by stress-optical measurements in the glass portion of a test seal (ISO 4790).

Apart from characteristic material values such as the coefficient of linear thermal expansion, transformation temperature, and elastic properties, the cooling rate (Fig. 3.4-31) and the shape can also have a considerable influence on the degree and distribution of seal stresses. The material combinations for sealing between metals and ceramics recommended for Schott glasses are shown in Fig. 3.4-32.

Types of Sealing Glasses

Sealing glasses may be classified by reference to the expansion coefficients of metals (e.g. tungsten and molybdenum) and alloys (Ni–Fe–Co, Ni–Fe–Cr, and other alloys) with which they are used. Hence sealing

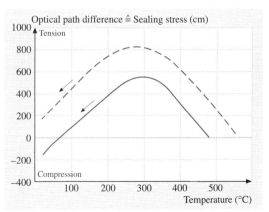

Fig. 3.4-31 Influence of the cooling rate on the sealing stress in an 8516–Ni/Fe combination. The *lower curve* corresponds to a low cooling rate; the *upper curve* corresponds to a high cooling rate

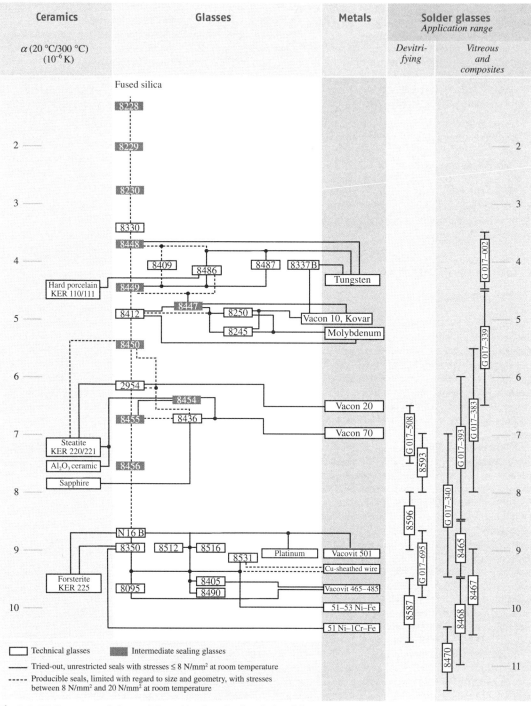

Fig. 3.4-32 Recommended material combinations for "graded seals"

Table 3.4-20 Special properties and principal applications of technically important sealing glasses, arranged according to their respective sealing partners

Metal $\alpha_{20/300}$ (10^{-6}/K)	Glass number	Glass characteristics	Principal applications as sealing glass
Tungsten (4.4)	8486	Alkaline-earth borosilicate, high chemical resistance, high working temperature. Suprax®	Lamp bulbs
	8487	High boron content, low melting temperature	Discharge lamps, surge diverters
Molybdenum (5.2)	8412	Alkaline-earth borosilicate, high chemical resistance. Fiolax® clear	Lamp bulbs
	8253	Alkaline-earth aluminosilicate glass	Lamp interior structures, lamp bulbs
Molybdenum and 28Ni/18Co/Fe (5.1)	8250	High boron content, low melting temperature, high electrical insulation, low dielectric losses	Transmitter tubes, image converters, TV receiver tubes
	8245	High boron content, low melting temperature, low X-ray absorption	X-ray tubes
28Ni/23Co/Fe (7.7)	8454	Alkali–alkaline-earth silicate, sealable with steatite and Al_2O_3 ceramics	Intermediate sealing glass
	8436	Alkali–alkaline-earth silicate, sealable with sapphire, resistant to Na vapor and alkalis	Special applications
51Ni/1Cr/Fe (10.2)	8350	Soda–lime silicate glass. AR glass	Tubes
Cu-sheathed wire ($\alpha_{20/400}$ radial 99, $\alpha_{20/400}$ axial 72)	8095	Alkali–lead silicate, high electrical insulation	Lead glass, stem glass for electric lamps and tubes
	8531	Dense lead silicate, Na- and Li-free, low melting temperature	Low-temperature encapsulation of diodes
	8532	High electrical insulation	
52–53Ni/Fe (10.2–10.5)	8512	Contains FeO for hot forming by IR, lead-free	Reed switches
	8516	Contains FeO for hot forming by IR, low volatilization, lead-free	Reed switches

glasses may be referred to as "tungsten sealing glasses", "Kovar glasses", etc. (see Table 3.4-20).

Alkaline-earth borosilicate glasses (8486 and 8412) and aluminosilicate glasses (8252 and 8253) have the necessary sealability and thermal resistance to be particularly suitable for the tungsten and molybdenum seals frequently used in heavy-duty lamps.

Ni–Fe–Co alloys, which often substitute for molybdenum, require that the transformation temperature be limited to 500 °C maximum. Suitable glasses (8250 and 8245) characteristically contain relatively high amounts of B_2O_3. These glasses have additional special properties, such as high electrical insulation, low dielectric loss, and low X-ray absorption, and meet the most stringent requirements for vacuum-tube technology and electronic applications.

For Ni–Fe–(Cr) alloys, which are frequently used in technological applications, as well as for copper-sheathed wire, glass groups belonging to the soft-glass category are recommended. Such glasses usually meet certain special requirements, such as high electrical insulation (e.g. alkali–lead silicate, 8095) or an exceptionally low working temperature (e.g. the dense-lead glasses 8531 and 8532).

FeO-containing glasses (8512 and 8516) are frequently used for hermetic encapsulation of electrical switches and electronic components in an inert gas. Hot forming and sealing are easily achieved by the absorption of IR radiation with an intensity maximum at 1.1 μm wavelength (Fig. 3.4-33). The presence of a proportion of Fe_2O_3 makes these glasses appear green. At appropriately high IR intensities, they require considerably shorter processing times than do flame-heated clear glasses.

Compression Seals

A common feature of all compression seals is that the coefficient of thermal expansion of the external metal part is considerably higher than the thermal expansion coefficients of the sealing glass and the metallic inner partner (conductor). As a result, the glass body is under overall radial pressure after sealing. This prestressing protects the glass body against dangerous mechanical loads. Because the compressive stress of the glass is compensated by a tensile stress in the jacket, the jacket wall must be sufficiently thick (at least 0.5 mm, even for small seals) to be able to permanently withstand such tension. If the thermal expansion of the metallic inner partner is lower than that of the sealing glass, an additional prestressing of the glass body results.

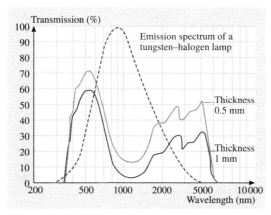

Fig. 3.4-33 IR absorption of Fe-doped glasses compared with the emission of a tungsten–halogen lamp at 3000 K (in relative units). The transmission of reed glass 8516 with thicknesses 0.5 mm and 1 mm is shown

Glasses for Sealing to Ceramics

Dielectrically superior, highly insulating ceramics such as hard porcelain, steatite, Al_2O_3 ceramics, and forsterite exhaust almost the complete expansion range offered by technical glasses. Hard porcelain can generally be sealed with alkaline-earth borosilicate glasses (for example 8486), which are also compatible with tungsten. Glass seals to Al_2O_3 ceramics and steatite are possible with special glasses such as 8454 and 8436, which will also seal to a 28Ni/18Co/Fe alloy. Soft glasses with thermal expansions around 9×10^{-6}/K are suitable for sealing to forsterite.

Intermediate Sealing Glasses

Glasses whose thermal expansion differs so widely from that of the partner component that direct sealing is impossible for reasons of stress must be sealed with intermediate sealing glasses. These glasses are designed in such a way that for the recommended combinations of glasses, the sealing stress does not exceed $20\,N/mm^2$ at room temperature (Table 3.4-21).

3.4.8.2 Solder and Passivation Glasses

Solder glasses are special glasses with a particularly low softening point. They are used to join glasses to other glasses, ceramics, or metals without thermally damaging the materials to be joined. Soldering is carried out in the viscosity range 10^4–10^6 dPa s of the solder glass; this corresponds to a temperature range $T_{solder} = 350$–$700\,°C$.

One must distinguish between vitreous solder glasses and devitrifying solder glasses, according to their behavior during the soldering process.

Vitreous solder glasses behave like traditional glasses. Their properties do not change during soldering; upon reheating of the solder joint, the temperature dependence of the softening is the same as in the preceding soldering process.

Unlike vitreous solder glasses, devitrifying solder glasses have an increased tendency to crystallize. They change into a ceramic-like polycrystalline state during soldering. Their viscosity increases by several orders of magnitude during crystallization so that further flow is suppressed. An example of this time-dependent viscosity behavior is shown in Fig. 3.4-34 for a devitrifying solder glass processed by a specific temperature–time program. Crystallization allows a stronger thermal reloading of the solder joint, up to the temperature range of the soldering process itself (e.g. glass 8596 has a soldering temperature of approx-

Table 3.4-21 Intermediate sealing glasses and the combinations of sealing partners in which they are used

Glass no.	Sealing partners[a]	$\alpha_{20/300}$ (10⁻⁶/K)	Transformation temperature T_g (°C)	Temperature at viscosity 10^{13} dPa s (°C)	$10^{7.6}$ dPa s (°C)	10^4 dPa s (°C)	Density ϱ (g/cm³)	T_{k100} (°C)
N16B	KER 250, Vacovit 501, Platinum – N16B 8456 (Red Line®)	8.8	540	540	720	1045	2.48	128
2954	KER 220, KER 221, Vacon 20 – 2954	6.3	600	604	790	1130	2.42	145
4210	Iron–4210	12.7	450	455	615	880	2.66	–
8228	Fused silica–8228–8229	1.3	∼700	726	1200	1705	2.15	355
8229	8228–8229–8230	2.0	630	637	930	1480	2.17	350
8230	8229–8230–8330	2.7	570	592	915	1520	2.19	257
8447	8412–8447–Vacon 10	4.8	480	505	720	1035	2.27	271
8448	8330–8448–8449, 8486, 8487	3.7	510	560	800	1205	2.25	263
8449	8486, 8487 – 8449 – 8447, 8412	4.5	535	550	785	1150	2.29	348
8450	8412–8450 – KER 220, 2954, 8436	5.4	570	575	778	1130	2.44	200
8454	KER 221, Al₂O₃ – 8454 – Vacon 70	6.4	565	575	750	1070	2.49	210
8455	2954, 8436, 8454 – 8455–8456	6.7	565	–	740	1030	2.44	–
8456	8455–8456 – N16B, 8350	7.4	445	–	685	1145	2.49	–

[a] Type designation of ceramics according to DIN 40685; manufacturer of Vacon alloys Vacuumschmelze Hanau (VAC).

imately 450 °C and a maximum reload temperature of approximately 435 °C).

The development of solder glasses (Table 3.4-22) with very low soldering temperatures is limited by the fact that reducing the temperature generally means increasing the coefficient of thermal expansion. This effect is less pronounced in devitrifying solder glasses. It can be avoided even more effectively by adding inert (nonreacting) fillers with low or negative coefficients of thermal expansion (for example ZrSiO₄ or β-eucryptite). The resulting glasses are called *composite solder glasses*. As a rule, the coefficient of thermal expansion of a solder glass should be smaller than the expansion coefficients of the sealing partners by $\Delta\alpha = 0.5 - 1.0 \times 10^{-6}$/K.

Up to their maximum service temperature, solder glasses are moisture- and gas-proof. Their good elec-

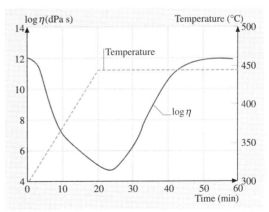

Fig. 3.4-34 Variation of the viscosity of a crystallizing solder glass during processing

Table 3.4-22 Schott solder glasses

Glass number	$\alpha_{(20/300)}$ (10^{-6}/K)	T_g (°C)	T at viscosity $10^{7.6}$ dPa s (°C)	Firing conditions T (°C)	Firing conditions t_{hold} (min)	Density ϱ (g/cm^3)	T_{k100} (°C)	ε_r	tan δ (10^{-4})
Vitreous and composite solder glasses									
G017-002[a]	3.6	540	650	700	15	3.4	–	6.8	37
G017-339[a]	4.7[b]	325	370	450	30	4.0	320	11.5	19
G017-383[a]	5.7[b]	325	370	430	15	4.7	325	13.0	15
G017-393[a]	6.5[b]	320	370	425	15	4.8	305	11.6	15
G017-340[a]	7.0[b]	315	360	420	15	4.8	320	13.4	14
8465	8.2	385	460	520	60	5.4	375	14.9	27
8467	9.1	355	420	490	60	5.7	360	15.4	29
8468	9.6	340	405	450	60	6.0	335	16.3	31
8470	10.0	440	570	680	60	2.8	295	7.7	15.5
8471	10.6[b]	330	395	440	30	6.2	–	17.1	52
8472	12.0[b]	310	360	410	30	6.7	–	18.2	47
8474	19.0[b]	325	410	480	30	2.6	170	7.2	5
Devitrifying solder glasses									
G017-508	6.5	365	–	530[c]	60	5.7	340	15.6	206
8593	7.7	300	–	520[c]	30	5.8	230	21.3	260
8596	8.7	320	–	450[c]	60	6.4	280	17.4	58
G017-695	8.9	310	–	425[c]	45	5.7	275	15.4	54
8587	10.0	315	–	435[c]	40	6.6	265	22.1	33

[a] Composite. [b] $\alpha_{20/250}$. [c] Heating rate 7–10 °C/min.

trical insulating properties are superior to those of many standard technical glasses. They are therefore also suitable as temperature-resistant insulators. The chemical resistance of solder glasses is generally lower than that of standard technical glasses. Therefore, solder glass seals can be exposed to chemically aggressive environments (e.g. acids or alkaline solutions) only for a limited time.

Passivation glasses are used for chemical and mechanical protection of semiconductor surfaces. They are generally zinc–borosilicate or lead–alumina–silicate glasses.

To avoid distortion and crack formation, the different coefficients of thermal expansion of the passivation glass and the semiconductor component must be taken into account. If the mismatch is too large, a network of cracks will originate in the glass layer during cooling or subsequent processing and destroy the hermetic protection of the semiconductor surface. There are three ways to overcome this problem:

- A thinner passivation glass layer. Schott recommends a maximum thickness for this layer.
- Slow cooling in the transformation range. As a rough rule, a cooling rate of 5 K/min is suitable for passivation layers in the temperature range $T_g \pm 50$ K.
- Use of composite glasses. Composites can be made with an inert filler such as a powdered ceramic with a very low or negative thermal expansion.

Properties of Passivation Glasses

The electrical insulation, including the dielectric breakdown resistance, generally depends on the alkali content, particularly the Na$^+$ content. Typical contents are below 100 ppm for Na$_2$O and K$_2$O, and below 20 ppm for Li$_2$O. Heavy metals which are incompatible with semiconductors are controlled as well. The CuO content, for example, is below 10 ppm.

Because the mobility of charge carriers increases drastically with increasing temperature, a temperature limit, called the junction temperature T_j, is defined up to which glass-passivated components can be used in blocking operations.

Various types of passivation glasses are listed in Table 3.4-23.

3.4.8.3 Colored Glasses

In physics, color is a phenomenon of addition or subtraction of parts of the visible spectrum, due to selective absorption or scattering in a material. The light transmission through a sample of thickness d at a wavelength λ is described by Lambert's law,

$$\tau_i(\lambda) = \exp\left[-\sum\sum \varepsilon_n(\lambda, c_n, c_m)d\right], \quad (4.42)$$

where ε is the extinction coefficient, which depends on the wavelength and the concentration of the active agents. For low concentrations, ε is additive and proportional to the concentration, and we obtain Beer's law,

$$\tau_i(\lambda) = \exp\left[-\sum \varepsilon_n(\lambda)c_n d\right], \quad (4.43)$$

where ε now depends only on the wavelength and the specific species or process n. In glasses, the extinction is caused by electronic and phononic processes in the UV and IR regions, respectively, and by absorption and scattering by ions, lattice defects, and colloids and microcrystals in the visible region. Different oxidation states of one atom, for example Fe^{2+} and Fe^{3+}, must be treated as different species. Charge transfer and ligand fields are examples of multiatom mechanisms that modify the absorption characteristics. The position of the maximum-extinction peak depends on the refractive index of the base glass; for example, for Ag metal colloids the position of the peak shifts from $\lambda_{max} = 403$ nm in Duran®, with $n_d = 1.47$, to $\lambda_{max} = 475$ nm in SF 56, with $n_d = 1.79$.

Table 3.4-23 Schott passivation glasses

Glass number	Type	Typical applications	$\alpha_{(20/300)}$ (10^{-6}/K)	T_g (°C)	Pb content (wt%)	Sealing temp. (°C)	Sealing time (min)	T_j (°C)	Layer thickness (μm)
G017-057	Zn–B–Si	Sintered glass diodes	4.5	546	1–5	690	10	180	–
G017-388	Zn–B–Si composite	Thyristors, high-blocking rectifiers	3.6	550	1–5	700	5	180	≤ 30
G017-953	Zn–B–Si composite		2.81	a	1–5	770	30	180	–
G017-058	Zn–B–Si	Sintered glass diodes	4.5	543	1–5	690	10	180	–
G017-002	Zn–B–Si composite	Sintered glass diodes	3.7	545	1–5	700	10	180	–
G017-984	Zn–B–Si	Stack diodes	4.6	538	5–10	720	10	180	–
G017-096R	Pb–B–Si	Sintered glass diodes, planar and mesa diodes	4.8	456	10–50	680	5	160	–
G017-004	Pb–B–Si	Mesa diodes	4.1	440	10–50	740	5	160	≤ 30
G017-230	Pb–B–Si composite	Power transistors	4.2	440	10–50	700	5	160	≤ 25
G017-725	Pb–B–Si	Sintered glass diodes	4.9	468	10–50	670	10	180	–
G017-997	Pb–B–Si composite	Wafers	4.4	485	10–50	760	20	180	–
G017-209	Pb–Zn–B	ICs, transistors	6.6	416	10–50	510	10	180	≤ 5
G017-980	Pb–Zn–B	Varistors			10–50			–	–
Vitreous			6.5	393		520	30	–	–
Devitrified			5.8	a		620	30	–	–
G018-088	Pb–Zn–B composite	Varistors	4.88	425	10–50	560	30	–	–

[a] Cannot be determined.

Colored glasses are thus technical or optical colorless glasses with the addition of coloring agents. A collection of data can be found in [4.17]. They are widely used as optical filters for various purposes, such as short-pass or long-pass edge filters, or in combinations of two or more elements as band-pass or blocking filters.

Owing to the absorbed energy, inhomogeneous heating occurs (between the front and rear sides, and, especially, in the radial direction), which results in internal stress via the thermal expansion under intense illumination. Precautions have to be taken in the mechanical mounting to avoid breakage. The application temperature T should satisfy the conditions $T \leq T_g - 300\,°C$ in the long term and $T \leq T_g - 250\,°C$ for short periods. Prestressing may be necessary to improve the breaking strength in heavy-load applications.

The Schott filter glasses are classified into groups listed in Table 3.4-24:

Table 3.4-24 Groups of Schott Filter glasses

UG	Black and blue glasses, ultraviolet-transmitting
BG	Blue, blue–green, and multiband glasses
VG	Green glasses
OG	Orange glasses, IR transmitting
RG	Red and black glasses, IR-transmitting
NG	Neutral glasses with uniform attenuation in the visible
WG	Colorless glasses with different cutoffs in the UV, which transmit in the visible and IR
KG	Virtually colorless glasses with high transmission in the visible and absorption in the IR (heat protection filters)
FG	Bluish and brownish color temperature conversion glasses

DIN has defined a nomenclature to allow one to see the main optical properties for a reference thickness d from the identification symbol (see Table 3.4-25):
Multiband filters and color conversion filters are not specified by DIN.

Ionically Colored Glasses

Ions of heavy metals or rare earths influence the color when in true solution. The nature, oxidation state, and quantity of the coloration substance, as well as the type of the base glass, determine the color (Figs. 3.4-35 – 3.4-38, and Table 3.4-26).

Table 3.4-25 DIN nomenclature for optical filter glasses

Band-pass filters	BP $\lambda_{max}/\Delta\lambda_{HW}$, where λ_{max} = wavelength of maximum internal transmission, and $\Delta\lambda_{HW}$ = bandwidth at 50% internal transmission.
Short-pass filters	KP $\lambda_{50\%}$, where $\lambda_{50\%}$ = cutoff wavelength at 50% internal transmission.
Long-pass filters	LP $\lambda_{50\%}$, where $\lambda_{50\%}$ = cutoff wavelength at 50% internal transmission.
Neutral-density filters	N τ, where τ = internal transmission at 546 nm.

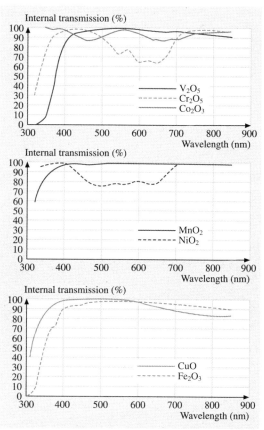

Fig. 3.4-35 Spectral internal transmission of BK7, colored with various oxides; sample thickness 100 mm

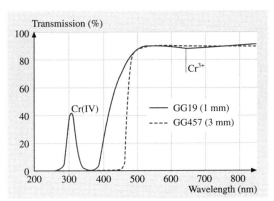

Fig. 3.4-36 Transmission spectra of yellow glasses

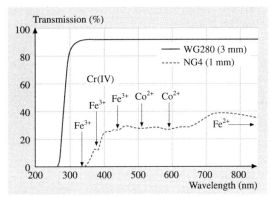

Fig. 3.4-38 Transmission spectra of gray and white glasses

Colloidally Colored Glasses

The colorants of these glasses are, in most cases, rendered effective by a secondary heat treatment (striking) of the initially nearly colorless glass. Particularly important glasses of this type are the yellow, orange, red, and black filter glasses, with their steep absorption edges. As with ionic coloration, the color depends on the type and concentration of the additives, on the type of the base glass, and on the thermal history, which determines the number and diameter of the precipitates (Fig. 3.4-39 and Table 3.4-27).

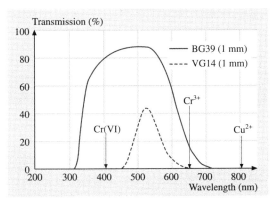

Fig. 3.4-37 Transmission spectra of blue and green glasses

Table 3.4-27 Colors of some metal colloids

Element	Peak position (nm)	n_d	Color
Ag	410	1.5	Yellow
Cu	530–560	?	Red
Au	550	1.55	Red
Se	500	?	Pink

Table 3.4-26 Colors of some ions in glasses

Element	Valency	Color
Fe	2+	Green, sometimes blue
Fe	3+	Yellowish brown
Cu	2+	Light blue, turquoise
Cr	3+	Green
Cr	6+	Yellow
Ni	2+	Violet (tetrahedral coordination)
Ni	2+	Yellow (octahedral coord.)
Co	2+	Intense blue
Co	3+	Green
Mn	2+	Pale yellow
Mn	3+	Violet
V	3+	Green (silicate), brown (borate)
Ti	3+	Violet (reducing melt)
Pr	3+	Light green
Nd	3+	Reddish violet
Er	3+	Pale red

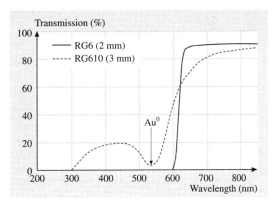

Fig. 3.4-39 Transmission spectra of red glasses

Doping with semiconductors results in microcrystalline precipitates which have band gap energies in the range 1.5–3.7 eV, corresponding to wavelengths in the range 827–335 nm. The preferred materials are ZnS, ZnSe, ZnTe, CdS, CdSe, and CdTe, which also form solid solutions. By mixing these dopants, any cutoff wavelength between 350 nm and 850 nm can be achieved (Table 3.4-30).

3.4.8.4 Infrared-Transmitting Glasses

The transmission in the infrared spectral region is limited by phonons and by local molecular vibrations and their overtones. The vibration frequencies decrease with increasing atomic mass. This intrinsic absorption has extrinsic absorption caused by impurities and lattice defects, such as hydroxyl ions, dissolved water, and microcrystals, superimposed on it. An overview can be found in [4.12].

Oxide Glasses

The transmission is determined by the vibrations of the common network formers [4.19] (Fig. 3.4-40). The vibrations of the network modifiers are found at longer wavelengths.

The heavy-metal oxide (HMO) glasses are transparent up to approximately 7 μm for a 1 mm thickness. But often they show a strong tendency toward devitrification, which very much limits the glass-forming compositions. The transmission of some commercial glasses is shown in Fig. 3.4-41.

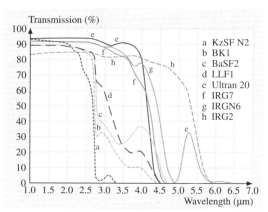

Fig. 3.4-41 Infrared transmission spectra of some Schott optical and special IR glasses; thickness 2.5 mm. The OH absorption of the glasses may vary owing to differences in the raw materials and the melting process

Halide Glasses

These glasses use F, Cl, Br, and I (halogens) as anions instead of oxygen. The transmission range is extended up to approximately 15 μm [4.20]. The oldest halide glasses are BeF_2, $ZnCl_2$, and AlF_3, which, however, have limited application owing to their toxicity, tendency toward crystallization, and hygroscopic behavior.

Some new glasses use ZrF_4, HfF_4, and ThF_4 as glass formers, and BaF_2, LaF_3 (heavy-metal flourides, HMFs) as modifiers. They are often named after their cation composition; for example, a glass with a cation composition $Zr_{55}Ba_{18}La_6Al_4Na_{17}$ would be called a ZBLAN glass.

Chalcogenide Glasses

These glasses use S, Se, and Te (chalcogens) as anions instead of oxygen. The transmission range is extended up to approximately 30 μm. Stable glass-forming regions are found in the Ge–As–S, Ge–As–Se, and Ge–Sb–Se systems; an example of a commercial glass is $Ge_{30}As_{15}Se_{55}$.

A combination of chalcogenides with halides is found in the TeX glasses, for example Te_3Cl_2S and Te_2BrSe.

The main properties of infrared-transmitting glasses are compiled in Tables 3.4-29 and 3.4-30.

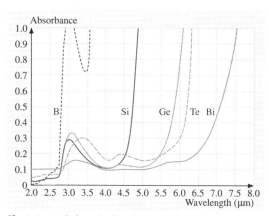

Fig. 3.4-40 Infrared absorbance of analogous oxide glasses with different network-forming cations [4.18]; sample thickness 1.85 mm, except for Bi, where the thickness was 2.0 mm

Table 3.4-28 Properties of Schott filter glasses

Glass type	DIN identification	Reference thickness d_r (mm)	Density ϱ (g/cm³)	Refractive index n_d	Chemical resistance Stain FR	Chemical resistance Acid SR	Chemical resistance Alkali AR	Transformation temperature T_g (°C)	Thermal expansion $\alpha_{-30/+70°C}$ (10^{-6}/K)	Thermal expansion $\alpha_{20/300°C}$ (10^{-6}/K)	Temp. coeff. T_K (nm/K)
UG1	BP 351/78	1	2.77	1.54	0	1.0	1.0	603	7.9	8.9	–
UG5	BP 318/173	1	2.85	1.54	0	3.0	2.0	462	8.1	9.4	–
UG11	BP 324/112 + BP 720/57	1	2.92	1.56		3.0	2.0	545	7.8	9	–
BG3	BP 378/185	1	2.56	1.51	0	1.0	1.0	478	8.8	10.2	–
BG4	BP 378/165	1	2.66	1.53		1.0	1.0	536	7.7	9	–
BG7	BP 466/182	1	2.61	1.52		1.0	1.0	468	8.5	19.9	–
BG12	BP 409/140	1	2.58	1.52		1.0	1.0	480	8.6	10.1	–
BG18	BP 480/250 + KP 605	1	2.68	1.54	0	2.0	2.0	459	7.4	8.8	–
BG20	Multiband	–	2.86	1.55	0	1.0	1.0	561	8.3	9.3	–
BG23	BP 459/232 + KP 575	1	2.37	1.52	0	1.0	1.0	483	8.9	10.2	–
BG24A	BP 342/253	1	2.72	1.53	0	3.0	1.0	460	8.5	9.7	–
BG25	BP 401/156	1	2.56	1.51	0	1.0	1.0	487	8.7	10.1	–
BG26	–	1	2.56	1.51		1.0	1.0	494	8.8	10.3	–
BG28	BP 436/156 + KP 514	1	2.60	1.52	0	1.0	1.0	474	8.7	10	–
BG34	Color conversion	2	3.23	1.59	0	1.0	1.0	441	9.9	10.7	–
BG36	Multiband	–	3.62	1.69	1	52.2	1.2	660	6.1	7.2	–
BG38	BP 487/334 + KP 654	1	2.62	1.53	0	2.0	2.0	466	7.5	8.9	–
BG39	BP 475/269 + KP 609	1	2.73	1.54	0	5.1	3.0	321	11.6	13.1	–
BG40	BP 482/318 + KP 641	1	2.67	1.53	0	5.1	3.0	305	11.9	13.7	–
BG42	BP 478/253 + KP 604	1	2.69	1.54	0	2.0	2.0	477	7.3	8.7	–
VG6	BP 523/160	1	2.90	1.55	0	1.0	1.0	470	9.1	10.6	–
VG9	BP 530/114	1	2.87	1.55	0	1.0	1.0	470	9.2	10.6	–
VG14	BP 52487	1	2.89	1.56	0	1.0	1.0	470	9.2	10.6	–
GG385	LP 385	3	3.22	1.58	0	2	2.3	459	7.7	8.8	0.07
GG395	LP 395	3	3.61	1.62	0	1	2.3	438	7.7	8.6	0.08
GG400	LP 400	3	2.75	1.54	3	4.4	1.0	595	9.6	10.5	0.07
GG420	LP 420	3	2.76	1.54	3	4.4	1.0	586	9.6	10.5	0.07

Table 3.4-28 Properties of Schott filter glasses, cont.

GG435	LP 435	3	2.75	1.54	3	4.4	1.0	605	9.5	10.5	0.07
GG455	LP 455	3	2.75	1.54	3	4.4	1.0	600	9.7	10.5	0.08
GG475	LP 475	3	2.75	1.54	3	4.4	1.0	594	9.8	10.6	0.09
GG495	LP 495	3	2.75	1.54	3	4.4	1.0	600	9.6	10.6	0.10
OG515	LP 515	3	2.76	1.54	3	4.4	1.0	597	9.7	10.6	0.11
OG530	LP 530	3	2.75	1.54	3	4.4	1.0	595	9.7	10.6	0.12
OG550	LP 550	3	2.75	1.54	3	4.4	1.0	597	9.6	10.7	0.13
OG570	LP 570	3	2.75	1.54	3	4.4	1.0	596	9.7	10.7	0.14
OG590	LP 590	3	2.75	1.54	3	4.4	1.0	599	9.8	10.6	0.15
RG9	BP 885307 + LP 731	3	2.76	1.54	3	4.4	1.0	581	9.8	10.7	0.07
RG610	LP 610	3	2.75	1.54	3	4.4	1.0	595	9.8	10.7	0.16
RG630	LP 630	3	2.76	1.54	3	4.4	1.0	597	9.6	10.7	0.17
RG645	LP 645	3	2.76	1.54	3	4.4	1.0	597	9.6	10.7	0.17
RG665	LP 665	3	2.75	1.54	3	4.4	1.0	592	9.8	10.8	0.17
RG695	LP 695	3	2.76	1.54	3	3	1.0	599	9.6	10.6	0.18
RG715	LP 715	3	2.75	1.54	3	3	1.0	589	9.8	10.7	0.18
RG780	LP 780	3	2.9	1.56	5	52.4	1.0	571	9.7	10.7	0.22
RG830	LP 830	3	2.94	1.56	5	53.4	1.0	569	9.5	10.5	0.23
RG850	LP 850	3	2.93	1.56	5	53.4	1.0	571	9.5	10.5	0.24
RG1000	LP 1000	3	2.75	1.55	0	1	1.2	478	9.2	9.9	0.38
NG1	N 10-4	1	2.49	1.52	1	2.2	1.0	466	6.6	7.2	–
NG3	N 0.09	1	2.44	1.51	1	2.2	1.0	462	6.5	7.3	–
NG4	N 0.27	1	2.43	1.51	1	3.2	2.0	483	6.7	7.2	–
NG5	N 0.54	1	2.43	1.5	1	3.2	2.0	474	6.6	7.3	–
NG9	N 0.04	1	2.44	1.51	1	3.2	2.0	487	6.4	7.3	–
NG10	N 0.004	1	2.47	1.52	1	3.2	2.0	468	6.4	7.2	–
NG11	N 0.72	1	2.42	1.5	1	3.4	2.0	473	6.9	7.5	–
NG12	N 0.89	1	2.34	1.49	4	51.4	2.0	460	5.9	6.4	–
WG225	LP 225	2	2.17	1.47	3	51.3	3.3	437	3.8	4.1	0.02
WG280	LP 280	2	2.51	1.52	0	1	2.0	563	7.0	8.3	0.04
WG295	LP 295	2	2.51	1.52	0	1	2.0	557	7.1	8.3	0.06
WG305	LP 305	2	2.59	1.52	0	1	1.0	546	8.2	9.6	0.06
WG320	LP 320	2	3.22	1.58	0	1	2.3	413	9.1	10.6	0.06
KG1	KP 751	2	2.53	1.52	0	2.0	3.0	599	5.3	6.1	–
KG2	KP 814	2	2.52	1.51	0	2.0	3.0	605	5.4	6.3	–
KG3	KP 708	2	2.52	1.51	0	2.0	4.0	581	5.3	6.1	–
KG4	KP 868	2	2.53	1.51	0	2.0	3.0	613	5.4	6.2	–
KG5	KP 689	2	2.53	1.51	0	3.0	4.0	565	5.4	6.2	–
FG3	Color conversion	2	2.37	1.5	0	1.0	1.0	564	5.4	6	–
FG13	Color conversion	2	2.78	1.56	1	3.4	1.0	556	9.6	10.5	–

Table 3.4-29 Property ranges of various infrared-transmitting glass types

Glass type	Transparency range[a] (μm)	Refractive index n_d	Abbe value ν_d	Thermal expansion α (10^{-6}/K)	Microhardness HV	Transformation temperature T_g (°C)	Density ϱ (g/cm^3)
Fused silica	0.17–3.7	1.46	67	0.51	800	1075	2.2
Oxide glasses	0.25–6.0	1.45–2.4	20–100	3–15	330–700	300–700	2–8
Fluorophosphate glasses	0.2–4.1	1.44–1.54	75–90	13–17	–	400–500	3–4
Fluoride glasses	0.3–8.0	1.44–1.60	30–105	7–21	225–360	200–490	2.5–6.5
Chalcogenide glasses	0.7–25	2.3–3.1[b]	105–185[c]	8–30	100–270	115–370	3.0–5.5

[a] 50% internal transmission for 5 mm path length.
[b] Infrared refractive index at 10 μm.
[c] Infrared Abbe value $\nu_{8-12} = (n_{10} - 1)/(n_8 - n_{12})$.

Table 3.4-30 Commercial infrared-transmitting glasses

Glass type	Glass name	Transparency range[a] (μm)	Refractive index n	Density ϱ (g/cm^3)	Thermal expansion α (10^{-6}/K)	Transformation temperature T_g (°C)	Manufacturer
Fused silica	SiO$_2$	0.17–3.7	1.4585[b]	2.20	0.51	1075	Corning, Heraeus, General Electric, Quartz et Silice, Schott Lithotec
Silicates	IRG7	0.32–3.8	1.5644[b]	3.06	9.6	413	Schott
	IRG15	0.28–4.0	1.5343[b]	2.80	9.3	522	Schott
	IRG3	0.40–4.3	1.8449[b]	4.47	8.1	787	Schott
Fluorophosphate	IRG9	0.36–4.2	1.4861[b]	3.63	16.1	421	Schott
Ca aluminate	9753[c]	0.40–4.3	1.597[d]	2.80	6.0	830	Corning
	IRG N6[c]	0.35–4.4	1.5892[b]	2.81	6.3	713	Schott
	WB37A[c]	0.38–4.7	1.669[b]	2.9	8.3	800	Sasson
	VIR6	0.35–5.0	1.601[b]	3.18	8.5	736	Corning France
	IRG11	0.38–5.1	1.6809[b]	3.12	8.2	800	Schott
	BS39B	0.40–5.1	1.676[b]	3.1	8.4	–	Sasson
Germanate	9754	0.36–5.0	1.664[b]	3.58	6.2	735	Corning
	VIR3	0.49–5.0	1.869[b]	5.5	7.7	490	Corning France
	IRG2	0.38–5.2	1.8918[b]	5.00	8.8	700	Schott
Heavy-metal oxide	EO	0.5–5.8	2.31[d]	8.2	11.1	320	Corning
Heavy-metal fluoride	ZBLA	0.30–7.0	1.5195[b]	4.54	16.8	320	Verre Fluoré
	Zirtrex	0.25–7.1	1.50[b]	4.3	17.2	260	Galileo
	HTF-1	0.22–8.1	1.517[d]	3.88	16.1	385	Ohara

Table 3.4-30 Commercial infrared-transmitting glasses, cont.

Glass type	Glass name	Transparency range[a] (μm)	Refractive index n	Density ϱ (g/cm^3)	Thermal expansion α (10^{-6}/K)	Transformation temperature T_g (°C)	Manufacturer
Chalcogenide	AMTIR1	0.8–12	2.5109[e]	4.4	12	362	Amorph. Materials
	IG1.1	0.8–12	2.4086[e]	3.32	24.6	150	Vitron
	AMTIR3	1.0–13	2.6173[e]	4.7	13.1	278	Amorph. Materials
	IG6	0.9–14	2.7907[e]	4.63	20.7	185	Vitron
	IG2	0.9–15	2.5098[e]	4.41	12.1	368	Vitron
	IG3	1.4–16	2.7993[e]	4.84	13.4	275	Vitron
	As$_2$S$_3$	1.0–16	2.653[e]	4.53	30	98	Corning France
	IG5	1.0–16	2.6187[e]	4.66	14.0	285	Vitron
	IG4	0.9–16	2.6183[e]	4.47	20.4	225	Vitron
	1173	0.9–16	2.616[e]	4.67	15	300	Texas Instruments

[a] 50% internal transmission for 5 mm path length.
[b] Refractive index at 0.587 μm.
[c] This glass contains SiO$_2$.
[d] Refractive index at 0.75 μm.
[e] Refractive index at 5 μm.

References

4.1 S. R. Elliott: *Physics of Amorphous Materials* (Longman, Harlow 1990)

4.2 H. Bach, D. Krause (Eds.): *Analysis of the Composition and Structure of Glass and Glass Ceramics* (Springer, Berlin, Heidelberg 1999) 2nd printing

4.3 H. Bach (Ed.): *Low Thermal Expansion Glass Ceramics* (Springer, Berlin, Heidelberg 1995)

4.4 W. Höland, G. Beall: *Glass Ceramic Technology* (American Ceramic Society, Westerville 2002)

4.5 N. P. Bansal, R. H. Doremus: *Handbook of Glass Properties* (Academic Press, Orlando 1986)

4.6 O. V. Mazurin, M. V. Streltsina, T. P. Shvaiko-Shvaikovskaya: *Handbook of Glass Data*, Part A, *Silica Glass and Binary Silicate Glasses*; Part B, *Single Component and Binary Non-silicate Oxide Glasses*; Part C, *Ternary Silicate Glasses*, Physical Sciences Data Series, Vol. 15, (Elsevier, Amsterdam 1983–1987)

4.7 R. Blachnik (Ed.): *Taschenbuch für Chemiker und Physiker*, D'Ans-Lax, Vol. 3, 4th edn. (Springer, Berlin, Heidelberg 1998)

4.8 *MDL SciGlass*, Version 4.0 (Elsevier, Amsterdam)

4.9 *INTERGLAD* (International Glass Database), Version 5, New Glass Forum

4.10 S. English, W. E. S. Turner: Relationship between chemical composition and the thermal expansion of glasses, J. Am. Ceram. Soc. **10**, 551 (1927); J. Am. Ceram. Soc. **12** (1929) 760

4.11 A. Paul: *Chemistry of Glasses* (Chapman and Hall, New York 1982)

4.12 H. Bach, N. Neuroth (Eds.): *The Properties of Optical Glass*, Schott Series on Glass and Glass Ceramics, 2nd printing (Springer, Berlin, Heidelberg 1998)

4.13 A. Winckelmann, F. O. Schott: Über thermische Widerstandskoeffizienten verschiedener Gläser in ihrer Abhängigkeit von der chemischen Zusammensetzung, Ann. Phys. (Leipzig) **51**, 730–746 (1894)

4.14 H. Scholze: *Glass* (Vieweg, Braunschweig 1965)

4.15 Schott: *Schott Technical Glasses* (Schott Glas, Mainz 2000)

4.16 H. R. Philipp: Silicon dioxide (SiO$_2$) (glass). In: *Handbook of Optical Constants of Solids*, ed. by E. D. Palik (Academic Press, New York 1985) pp. 749–763

4.17 C. R. Bamford: *Colour Generation and Control in Glass* (Elsevier, Amsterdam 1977)

4.18 W. H. Dumbaugh: Infrared-transmitting oxide glasses, Proc. SPIE **618**, 160–164 (1986)

4.19 N. Neuroth: Zusammenstellung der Infrarotspektren von Glasbildnern und Gläsern, Glastechn. Ber. **41**, 243–253 (1968)

4.20 J. Lucas, J.-J. Adam: Halide glasses and their optical properties, Glastechn. Ber. **62**, 422–440 (1989) W. Vogel: *Glass Chemistry* (Springer, Berlin, Heidelberg 1994)

Subject Index

π-bonded chain geometry 993
π-bonded chain model
– diamond(111)2×1 1005
$\alpha = 0$ glass-ceramics 558
$(CH_3NHCH_2COOH)_3 \cdot CaCl_2$ family 932
5CB
– liquid crystals 948
8CB
– liquid crystals 948
8OCB
– liquid crystals 949

A

Abbe value
– glasses 543, 548
Abrikosov vortices 717
absorption and fluorescence spectra of CdSe 1039
absorption coefficient
– two-photon 826
absorption spectra of spherical particles 1046
Ac actinium 84
acceptor surface level 1023
acceptor surface state 1022
accumulation layer 1020
accuracy 4
acids
– liquid crystals 946
acoustic band 1012
acoustic surface wave 906
acronyms
– solid surface 1026
actinium Ac
– elements 84
adatom 995
adopted numerical values for selected quantities 23
Ag silver 65
Ag-based materials 344
age hardening 198
AISI (American Iron and Steel Institute) 221
Al aluminium 78
Al bronzes 298
alkali aluminium silicates
– electrical properties 434
– mechanical properties 434
– thermal properties 434

alkali halides
– surface phonon energy 1017
alkali–alkaline-earth silicate glasses 530
alkali–lead silicate glasses 530
alkaline-earth aluminium silicates
– electrical properties 436
– glasses 530
– mechanical properties 435
– thermal properties 436
allotropic and high-pressure modifications
– elements 46
alloy
– cast irons 270
– cobalt 272
– elinvar 780
– invar 780
– lead, battery grid 413
– lead–antimony 412
– lead–tin 415
– magnesium 163
– Ti_3Al-based 210
– TiAl-based 213
– titanium 206
– wear resistant 274
alloy systems 296
Allred 46
Alnico 798
Al–O–N ceramics
– dielectric properties 447
– optical properties 447
alum family 911
– ferroelectrics 911
alumina
– electrical properties 446
– mechanical properties 445
– properties 445
– thermal properties 446
aluminium Al
– aluminium alloys 171
– aluminium production 171
– chemical properties 172
– cold working 195
– corrosion behavior 204
– elements 78
– hot working 195
– mechanical properties 172
– mechanical treatment 195
– surface layers 204
– work hardening 195

aluminium alloy
– abrasion resistance 190
– aging 198
– behavior in magnetic fields 194
– binary Al-based systems 174
– classification of aluminium alloys 179
– coefficient of thermal expansion 192
– creep behavior 187
– elastic properties 194
– electrical conductivity 194
– hardness 186
– homogenization 198
– machinability 192
– mechanical properties 180, 182
– nuclear properties 194
– optical properties 194
– physical properties 192, 193
– sheet formability 190
– soft annealing 197
– specific heat 194
– stabilization 197
– stress-relieving 198
– structure 182
– technical property 186
– technological properties 190
– tensile strength 186
– thermal softening 195
– work-hardenable 180
aluminium antimonide
– crystal structure, mechanical and thermal properties 610
– electromagnetic and optical properties 619
– electronic properties 616
– transport properties 618
aluminium arsenide
– crystal structure, mechanical and thermal properties 610
– electromagnetic and optical properties 619
– electronic properties 616
– transport properties 618
aluminium casting alloys
– mechanical properties 184
– structure 184
aluminium compounds
– crystal structure, mechanical and thermal properties 610

– electromagnetic and optical
  properties 619
– electronic properties 614
– mechanical properties 610
– phonon dispersion curves 612
– thermal conductivity 618
– thermal properties 610
– transport properties 617
aluminium nitride
– crystal structure, mechanical and
  thermal properties 610
– electromagnetic and optical
  properties 619
– electronic properties 616
– transport properties 618
aluminium phase diagram
– aluminium alloy phase diagram
  174
aluminium phosphide
– crystal structure, mechanical and
  thermal properties 610
– electromagnetic and optical
  properties 619
– electronic properties 616
– transport properties 618
aluminothermy 174
Al–Ni phase diagram 285
Am americium 151
americium Am
– elements 151
amorphous alloys
– cobalt–based 774
– iron–based 773
– nickel–based 774
amorphous materials 27, 39
amorphous metallic alloys 772
amount of substance
– definition 14
ampere
– SI base unit 14
amphiphilic compound 941
amphiphilic liquid crystal 942
anisotropic magnetoresistance
  1058
annealing coefficient
– glasses 548
annealing of steel 223
antiferroelectric crystal 903
antiferroelectric hysteresis loop 904
antiferroelectric liquid crystal 911
antiferroelectrics
– definition 903
– dielectric properties 903
– elastic properties 903
– pyroelectric properties 903
antimony Sb

– elements 98
aperiodic crystals 27
aperiodic materials 33
apparent tilt angle 935
Ar argon 128
area of surface primitive cell
– crystallographic formulas 986
argon Ar
– elements 128
arsenic As
– elements 98
ARUPS 997
As arsenic 98
astatine At
– elements 118
ASTM 241, 242
ASTM (American Society for Testing
  and Materials) 330
ASW 906
At astatine 118
atomic moment 755
atomic number Z
– elements 45
atomic radius
– elements 46
atomic scattering 1019
atomic, ionic, and molecular
  properties
– elements 46
atomically clean crystalline surface
  979
atom–surface potential
– surface phonons 1019–1020
Au gold 65
austenitizing 224

B

B boron 78
Ba barium 68
back-bond state 1006
bainite 223
BaMnF$_4$ family 922
band bending
– solid surfaces 1023
band gap see energy gap 592
band pass filters
– glasses 566
band structure
– aluminium compounds 614
– beryllium compounds 653
– boron compounds 606
– cadmium compounds 679
– group IV semiconductors and
  IV–IV compounds 589–592
– indium compounds 643

– magnesium compounds 657
– mercury compounds 688
– oxides of Ca, Sr, and Ba 662
– zinc compounds 668
barium Ba
– elements 68
barium oxide
– crystal structure, mechanical and
  thermal properties 660
– electromagnetic and optical
  properties 664
– electronic properties 661
– transport properties 663
barium titanate 915
base quantities 12
– ISO 13
base unit
– SI 13
basis
– crystal structure 28
BaTiO$_3$ 915
bcc positions
– surface diagrams 982
Be beryllium 68
becquerel
– SI unit of activity 19
benzene 946
berkelium Bk
– elements 151
beryllium Be
– elements 68
beryllium compounds 652
– crystal structure, mechanical and
  thermal properties 652
– electromagnetic and optical
  properties 655
– electronic properties 653
– mechanical and thermal properties
  652
– optical properties 655
– thermal properties 652
– transport properties 655
beryllium oxide 447
– crystal structure, mechanical and
  thermal properties 652
– electrical properties 448
– electronic properties 653
– mechanical properties 447
– thermal properties 448
beryllium selenide
– crystal structure, mechanical and
  thermal properties 652
– electronic properties 653
beryllium sulfide
– crystal structure, mechanical and
  thermal properties 652

– electronic properties 653
beryllium telluride
– crystal structure, mechanical and thermal properties 652
– electronic properties 653
Bethe–Slater–Pauling relation 755, 756
Bh Bohrium 124
Bi bismuth 98
biaxial crystals 826
binding energy 998
– metal 999
Bioverit
– glasses 559
BIPM (Bureau International des Poids et Mesures) 3, 11, 12
bismuth Bi
– elements 98
Bi − Sr − Ca − Cu − O (BSCCO) 736
Bi−Sr−Ca−Cu−O
– coherence lengths 736
– London penetration depths 736
– maximum T_c 736
– structural data 736
– superconducting properties 741
– upper critical fields 736
BK 7
– glasses 537
Bk berkelium 151
blue phase 942
BNN
– ferroelectric material 920
Bohrium Bh
– elements 124
boiling temperature
– elements 47
Bondi 46
boracite-type family 911, 921
borides
– physical properties 452
Borofloat
– glasses 528, 529
boron antimonide
– crystal structure, mechanical, and thermal properties 604
– electromagnetic and optical properties 610
– electronic properties 607
– transport properties 608
boron arsenide
– crystal structure, mechanical, and thermal properties 604
– electromagnetic and optical properties 610
– electronic properties 606

– transport properties 608
boron B
– elements 78
boron compounds
– crystal structure, mechanical and thermal properties 604
– electromagnetic and properties 610
– electronic properties 606
– mechanical properties 604
– thermal properties 604
– transport properties 608
boron nitride
– crystal structure, mechanical, and thermal properties 604
– electromagnetic and optical properties 610
– electronic properties 606
– transport properties 608
boron phosphide
– crystal structure, mechanical, and thermal properties 604
– electromagnetic and optical properties 610
– electronic properties 606
– transport properties 608
borosilicate glasses 529, 530
Br bromine 118
Bragg equation 40
brasses 298
Bravais cell 979
– 2D lattice 980
Bravais lattice 32
– elements 47
breathing-mode acoustic oscillations 1040
Brillouin scattering 906
Brillouin zone 913
– aluminium compounds 614
– beryllium compounds 653
– boron compounds 606
– cadmium compounds 678
– gallium compounds 626
– group IV semiconductors and IV–IV compounds 589
– indium compounds 643
– magnesium compounds 657
– mercury compounds 688
– oxides of Ca, Sr, and Ba 661
– zinc compounds 668
Brillouin zone corner 921
bromine Br
– elements 118
bronzes 298
BSCCO
– films 738

– single crystal 739
– tapes 739
– wires 739
buckled dimer 991
bulk electron density 998
bulk glassy alloys 217, 218
bulk mobility 1026
bulk modulus
– elements 47

C

C carbon 88
Ca calcium 68
cadmium Cd
– elements 73
cadmium compounds
– crystal structure, mechanical and thermal properties 676
– electromagnetic and optical properties 683
– electronic properties 678
– mechanical and thermal properties 676
– optical properties 683
– thermal properties 676
– transport properties 682
cadmium oxide
– crystal structure, mechanical and thermal properties 676
– electromagnetic and optical properties 683
– electronic properties 678
– transport properties 682
cadmium selenide
– crystal structure, mechanical and thermal properties 676
– electromagnetic and optical properties 683
– electronic properties 678
– transport properties 682
cadmium sulfide
– crystal structure, mechanical and thermal properties 676
– electromagnetic and optical properties 683
– electronic properties 678
– transport properties 682
cadmium telluride
– crystal structure, mechanical and thermal properties 676
– electromagnetic and optical properties 683
– electronic properties 678
– transport properties 682
calamitic liquid crystal 941, 942

calcium Ca
– elements   68
calcium oxide
– crystal structure, mechanical and thermal properties   660
– electromagnetic and optical properties   664
– electronic properties   661
– transport properties   663
californium Cf
– elements   151
candela
– SI base unit   15
capacitor   903
capillary viscometer   943
carat   23
carbide
– cemented   277
– electrical properties   466
– mechanical properties   466
– physical properties   458
– thermal properties   466
carbon C
– elements   88
carbon equivalent (CE)   268
carbon fibers
– physical properties   477
carbon steels   230
carrier concentration n_i
– gallium compounds   631
– indium compounds   647
cast
– classification   268
cast iron   268
– grades   268
– mechanical properties   270
casting technology   170
catalysis   1020
Cd cadmium   73
Ce cerium   142
cell surface   943
cellulose   481, 509
– cellulose acetate (CA)   509, 510
– cellulose acetobutyrate (CAB)   509, 510
– cellulose propionate (CP)   509, 510
– ethylcellulose (EC)   509, 510
– polymers   509
– vulcanized fiber (VF)   509, 510
cement   432
cemented carbides   277
centering types
– crystal structure   28
centrosymmetric media   1007
Cerabone

– glasses   559
ceramic capacitor   906
ceramic thin film   914
– ferroelectrics   903
ceramics   345, 431
– Al–O–N   447
– applications   432
– non-oxide   451
– oxide   437
– properties   432
– refractory   437
– silicon   433
– technical   437
– traditional   432
Ceran®
– glasses   559
– linear thermal expansion   558
Ceravital
– glasses   559
cerium Ce
– elements   142
cesium Cs
– elements   59
Cf californium   151
CGPM (Conférence Générale des Poids et Mesures)   11, 12
CGS
– electromagnetic system   21
– electrostatic system   21
– Gaussian system   21
cgs definitions of magnetic susceptibility   48
chalcogenide glasses   568, 571
channel conductivity   1020
characterization of optical glasses   547
Charpy impact strength   478
– polymers   478
chemical disorder   38
chemical stability
– glasses   530, 531
– optical glasses   550
chemical symbols
– element   45
chemisorption   1020
chiral smectic C
– liquid crystals   944
chlorine Cl
– elements   118
cholesteric phase   942
cholesteryl (cholest-5-ene) substituted mesogens
– liquid crystals   968
cholesteryl compound   944
chromium Cr
– elements   114

CIP (current in the plane of layer)   1051
CIPM (Comité International des Poids et Mesures)   3, 11
Cl chlorine   118
clamped crystal   907, 921, 934
clamped dielectric constant   907
classifications of liquid crystals   942
climatic influences
– glasses   550
cluster boundaries   907
cluster formation   907
CM
– liquid crystals   970
Cm curium   151
CMOS (complementary metal-oxide-semiconductor)   1059
Co cobalt   135
$Co_{17}RE_2$ 805
Co_5RE 805
coating   943
cobalt
– alloys   272
– applications   273
– hard-facing alloy   274
– mechanical properties   277
– superalloys   274
– surgical implant alloys   277
cobalt Co
– elements   135
cobalt corrosion-resistant alloys   276
cobalt-based corrosion-resistant alloys   276
CODATA (Committee on Data for Science and Technology)   4
coefficient of expansion   478
– polymers   478
coefficient of thermal expansion
– glasses   526
coercive field   904
coherence length   920
coherent phonon and Raman spectra   1041
colloidal synthesis
– nanostructured materials   1065
colored glasses
– colorants   567
– glasses   565
columnar phase   941
commensurate reconstruction   986
commercially pure titanium (cp-Ti)   206
communication technology   912

compacted (vermicular) graphite (CG) 268
complex perovskite-type oxide 909
complex refractive index
– zinc compounds 674
composite medium 1045
– dielectric constant 1045
composite solder glasses
– glasses 563
composite structures
– crystallography 34
compound semiconductor 1003
compressibility 478
– polymers 478
compression modulus
– elements 47
condensed matter 27
– classification 28
conductivity
– frequency-dependent 823
conductivity tensor
– elements 47
conductor
– nanoparticle-based 1043
confined electronic systems
– nanostructured materials 1035
confinement effect
– nanostructured materials 1031
constants
– fundamental 3
container glasses 529
continuous distribution of states 1022
continuous-cooling-transformation (CCT) diagram 238
controlled rolling 240
Convention du Mètre 12
conventional system
– ISO 13
conversion factor 945
– density 945
– diamagnetic anisotropy 945
– dipole moment 945
– dynamic viscosity 945
– kinematic viscosity 945
– molar mass 945
– temperatures of phase transitions 945
– thermal conductivity 945
cooperative interactions 903
coordination number
– elements 46, 47
copolymers
– physical properties 477, 483
copper
– unalloyed 297

copper alloys 296, 297
copper Cu
– elements 65
copper–nickel 300
copper–nickel–zinc 300
Co–RE
– phase equilibria 805
corrosion 1020
– resistance 218
Coulomb blockade
– nanostructured materials 1031, 1044
coupled plasmon modes 1048
Co–Sm 803
CPP (current flows perpendicular to the plane of the layer) 1051, 1054
Cr chromium 114
creep modulus 478
– polymers 478
critical field 904
critical slowing-down 907, 908
critical temperature
– elements 47, 48
– Pb alloys 699
CrNi steels 252
Cronstedt, swedish mineralogist 1032
crystal axes 980
crystal morphology 30
crystal optics 824
crystal structure
– beryllium compounds 652
– cadmium compounds 676
– group IV semiconductors 578
– III–V compounds 610
– indium compounds 638
– IV–IV compound semiconductors 578
– magnesium compounds 655
– mercury compounds 686
– oxides of Ca, Sr, and Ba 660
– zinc compounds 665
crystal structure, mechanical and thermal properties
– III–V compounds 621, 638
– III–V semiconductors 604
– II–VI compounds 652, 655, 660, 665, 676, 686
crystal symmetry
– elements 47
– ferroelectrics 915
crystalline ferroelectric 911
crystalline materials
– definition 27
crystalline surface
– atomically clean 979

crystallization
– glasses 524
crystallographic formulas 986
crystallographic properties
– elements 46, 47
crystallographic space group 946–952, 955–961, 964, 966, 968–971
crystallographic structure
– methods to investigate 39
crystallography
– concepts and terms 27
– rudiments of 27
crystals
– biaxial 826
– cubic 826
– isotropic 826
– uniaxial 826
Cryston
– glasses 559
Cs cesium 59
CS2004
– liquid crystals 975
Cu copper 65
cubic
– dielectrics 828, 838
cubic $BaTiO_3$ 916
cubic boron nitride 451
cubic crystal 47
cubic system 986
Curie point 906
Curie temperature 906
– surface 1009
Curie–Weiss constant 931, 932
curium Cm
– elements 151
current/voltage characteristics of films 1044
Cu–Ni
– electrical conductivity 302
– thermal conductivity 303
Cu–Zn
– electrical conductivity 302
– thermal conductivity 303
Cu–Zn phase diagram 299
cyclohexane 946
cycloolefine copolymer (COC) 486, 487
cyclosilicate 433

D

D deuterium 54
damping constant
– optical mode frequency 916
dangling bonds (DBs) 991

DAS 991
data storage media
– nanostructured materials 1031
Db dubnium 105
de Broglie wavelength 1035
Debye length
– solid surfaces 1020
Debye temperature Θ_D
– boron compounds 605
– cadmium compounds 678
– gallium compounds 623
– group IV semiconductors 584
– indium compounds 640
– IV–IV compound semiconductors 584
– mercury compounds 687
– metal surfaces 1013
– oxides of Ca, Sr, and Ba 661
– solid surfaces 1014
– surface phonons 1012
– zinc compounds 667
decimal multiples of SI units 19
degree Celsius
– unit of temperature 14
density ϱ 478, 945–954, 956–971
– aluminium compounds 611
– beryllium compounds 653
– boron compounds 604
– cadmium compounds 676
– elements 47
– gallium compounds 621
– group IV semiconductors and IV–IV compounds 579
– indium compounds 638
– magnesium compounds 656
– mercury compounds 686
– oxides of Ca, Sr, and Ba 660
– polymers 478
– temperature dependence 943
– zinc compounds 665
density of electronic states
– magnesium compounds 658
density of electronic states see also density of states 658
density of phonon states see also density of states 658
density of states (DOS)
– nanostructure 1034
dentistry 330
depletion layer 1020
– solid surfaces 1024
derived quantities 12
– ISO 13
derived units
– SI 16
– special names and symbols 16

deuteration 925
deuterium D
– elements 54
devices
– optical 903
– piezoelectric 903
– pyroelectric 903
devil's staircase 932
devitrifying solder glasses 562
DFB (distributed feedback) 1038
DFG (difference frequency generation) 825
diamagnetic anisotropy 943, 945
diamond
– crystal structure, mechanical and thermal properties 578–588
– electromagnetic and optical properties 601
– electronic properties 589–594
– transport properties 595
diamond positions
– surface diagrams 983
diamond-like structure 979
Dicor
– glasses 559
dielectric
– constant (low-frequency) 823
– dissipation factor 823
– elasticity 823
– general properties 822
– lossy 823
– low-frequency materials 822
– stiffness constant 823
dielectric anisotropy 943, 946, 947
dielectric anomaly 922, 934
dielectric constant ε 907, 947–958, 960–966, 968–971, 973, 1045
– aluminium compounds 619
– beryllium compounds 655
– boron compounds 610
– cadmium compounds 683
– elements 46
– gallium compounds 635, 637
– group IV semiconductors and IV–IV compounds 601–603
– indium compounds 650
– magnesium compounds 660
– mercury compounds 691
– oxides of Ca, Sr, and Ba 664
– zinc compounds 672
dielectric dispersion 907
dielectric dissipation factor
– glasses 538
dielectric function 1001
– surface layer 1007
dielectric loss tan δ

– gallium compounds 635
dielectric material
– properties 826
dielectric polarization 825
dielectric properties
– glasses 538
dielectric strength
– glasses 539
dielectric tensor 827
dielectrics
– α-iodic acid, α-HIO$_3$ 868
– α-mercuric sulfide, α-HgS 846
– α-silicon carbide, SiC 840
– α-silicon dioxide, α-SiO$_2$ 846
– α-zinc sulfide, α-ZnS 842
– β-barium borate, β-BaB$_2$O$_4$ 848
– 2-cyclooctylamino-5-nitropyridine, C$_{13}$H$_{19}$N$_3$O$_2$ 874
– 3-methyl 4-nitropyridine 1-oxide, C$_6$N$_2$O$_3$H$_6$ 868
– 3-nitrobenzenamine, C$_6$H$_4$(NO$_2$)NH$_2$ 878
– 4-(N1N-dimethylamino)-3-acetamidonitrobenzene 884
– ADA 856
– ADP 856
– aluminium oxide, α-Al$_2$O$_3$ 844
– aluminium phosphate AlPO$_4$ 844
– ammonium dideuterium phosphate, ND$_4$D$_2$PO$_4$ 856
– ammonium dihydrogen arsenate, NH$_4$H$_2$AsO$_4$ 856
– ammonium dihydrogen phosphate, NH$_4$H$_2$PO$_4$ 856
– ammonium Rochelle salt 870
– ammonium sulfate, (NH$_4$)$_2$SO$_4$ 872
– "banana" 872
– barium fluoride, BaF$_2$ 828
– barium formate, Ba(COOH)$_2$ 866
– barium magnesium fluoride, BaMgF$_4$ 872
– barium nitrite monohydrate, Ba(NO$_2$)$_2 \cdot$ H$_2$O 842
– barium sodium niobate, Ba$_2$NaNb$_5$O$_{15}$ 872
– Barium titanate, BaTiO$_3$ 864
– BBO 848
– berlinite 844
– beryllium oxide, BeO 840
– BGO 838
– BIBO 884
– bismuth germanium oxide, Bi$_{12}$GeO$_{20}$ 838

- bismuth silicon oxide, $Bi_{12}SiO_{20}$ 838
- bismuth triborate, BiB_3O_6 884
- BK7 Schott glass 828
- BMF 872
- BSO 838
- cadmium germanium arsenide, $CdGeAs_2$ 856
- cadmium germanium phosphide, $CdGeP_2$ 856
- cadmium selenide, CdSe 840
- cadmium sulfide, CdS 840
- cadmium telluride, CdTe 834
- calcite, $CaCO_3$ 844
- calcium fluoride, CaF_2 828
- calcium tartrate tetrahydrate, $Ca(C_4H_4O_6) \cdot 4H_2O$ 874
- CBO 868
- CDA 858
- cesium dideuterium arsenate, CsD_2AsO_4 856
- cesium dihydrogen arsenate, CsH_2AsO_4 858
- cesium lithium borate, $CsLiB_6O_{10}$ 858
- cesium triborate, CsB_3O_5 868
- cinnabar 846
- CLBO 858
- CNB 874, 875
- COANP 874
- copper bromide, CuBr 834
- copper chloride, CuCl 834
- copper gallium selenide, $CuGaSe_2$ 858
- copper gallium sulfide, $CuGaS_2$ 858
- copper iodide, CuI 834
- cubic $m3m$ (O_h) 828
- cubic $\bar{4}3m$ (T_d) 834
- cubic, 23 (T) 838
- D(+)-saccharose, $C_{12}H_{22}O_{11}$ 888
- DADP 856
- DAN 884
- DCDA 856
- deuterated L-arginine phosphate, $(ND_xH_{2-x})_2^+(CND)(CH_2)_3CH(ND_yH_{3-y})^+COO^- \cdot D_2PO_4^- \cdot D_2O$ 884
- diamond, C 830
- dipotassium tartrate hemihydrate, $K_2C_4H_4O_6 \cdot 0.5H_2O$ 886
- DKDA 858
- DKDP 858
- DKT 886
- DLAP 884, 886
- DRDA 860
- DRDP 860
- fluorite 828
- fluorspar 828
- forsterite 866
- gadolinium molybdate, $Gd_2(MoO_4)_3$ 876
- gallium antimonide, GaSb 834
- gallium arsenide, GaAs 834
- gallium nitride, GaN 840
- gallium phosphide, GaP 834
- gallium selenide, GaSe 838
- gallium sulfide, GaS 838
- germanium, Ge 830
- GMO 876
- greenockite 840
- halite 832
- hexagonal, 6 (C_6) 842
- hexagonal, $6mm$ (C_{6v}) 840
- hexagonal, $\bar{6}m2$ (D_{3h}) 838
- high-frequency (optical) properties 817
- Iceland spar 844
- indium antimonide, InSb 836
- indium arsenide, InAs 836
- indium phosphide, InP 836
- Irtran-3 828
- Irtran-6 834
- isotropic 828
- KB5 880
- KBBF 846
- KDA 860
- KDP 860
- KLINBO 864
- KTA 880
- KTP 880
- LBO 878
- L-CTT 874
- lead molybdate, $PbMoO_4$ 854
- lead titanate, $PbTiO_3$ 864
- LFM 876
- list of described substances 818
- lithium fluoride, LiF 830
- lithium formate monohydrate, $LiCOOH \cdot H_2O$ 876
- lithium gallium oxide, $LiGaO_2$ 876
- lithium iodate, α-$LiIO_3$ 842
- lithium metagallate 876
- lithium niobate (5% MgO-doped), $MgO:LiNbO_3$ 848
- lithium niobate, $LiNbO_3$ 848
- lithium sulfate monohydrate, $Li_2SO_4 \cdot H_2O$ 886
- lithium tantalate, $LiTaO_3$ 848
- lithium tetraborate, $Li_2B_4O_7$ 864
- lithium triborate, LiB_3O_5 878
- low-frequency properties 817
- magnesium fluoride, MgF_2 852
- magnesium oxide, MgO 830
- magnesium silicate, Mg_2SiO_4 866
- Maxwell's equations 824
- m-chloronitrobenzene, $ClC_6H_4NO_2$ 874, 875
- mNA 878
- m-nitroaniline 878
- MNMA 874
- monoclinic, 2 (C_2) 884
- N,2-dimethyl-4-nitrobenzenamine, $C_8H_{10}N_2O_2$ 874
- N-[2-(dimethylamino)-5-nitrophenyl]-acetamide 884
- nantokite 834
- numerical data 818, 828
- orthorhombic, 222 (D_2) 866
- orthorhombic, $mm2$ (C_{2v}) 872
- orthorhombic, mmm (D_{2h}) 866
- paratellurite 854
- physical properties 817
- PMMA (Plexiglas) 828
- POM 868
- potassium acid phthalate, $KH(C_8H_4O_4)$ 878
- potassium bromide, KBr 830
- potassium chloride, KCl 830
- potassium dideuterium arsenate, KD_2AsO_4 858
- potassium dideuterium phosphate, KD_2PO_4 858
- potassium dihydrogen arsenate, KH_2AsO_4 860
- potassium dihydrogen phosphate, KH_2PO_4 860
- potassium fluoroboratoberyllate, $KBe_2BO_3F_2$ 846
- potassium iodide, KI 830
- potassium lithium niobate, $K_3Li_2Nb_5O_{15}$ 864
- potassium niobate, $KNbO_3$ 880
- potassium pentaborate tetrahydrate, $KB_5O_8 \cdot 4H_2O$ 880
- potassium sodium tartrate tetrahydrate, $KNa(C_4H_4O_6) \cdot 4H_2O$ 870
- potassium titanate (titanyl) phosphate, $KTiOPO_4$ 880
- potassium titanyl arsenate, $KTiOAsO_4$ 880
- proustite 850
- pyrargyrite 850
- quartz 846

- RDA 860
- RDP 862
- Rochelle salt 870
- rock salt 832
- RTP 882
- rubidium dideuterium arsenate, RbD_2AsO_4 860
- rubidium dideuterium phosphate, RbD_2PO_4 860
- rubidium dihydrogen arsenate, RbH_2AsO_4 860
- rubidium dihydrogen phosphate, RbH_2PO_4 862
- rubidium titanate (titanyl) phosphate, $RbTiOPO_4$ 882
- rutile 852
- sapphire 844
- Silicon dioxide, SiO_2 828
- silicon, Si 830
- silver antimony sulfide, Ag_3SbS_3 850
- silver arsenic sulfide, Ag_3AsS_3 850
- silver gallium selenide, $AgGaSe_2$ 862
- silver thiogallate, $AgGaS_2$ 862
- sodium ammonium tartrate tetrahydrate, $Na(NH_4)C_4H_4O_6 \cdot 4H_2O$ 870
- sodium chlorate, $NaClO_3$ 838
- sodium chloride, NaCl 832
- sodium fluoride, NaF 832
- sodium nitrite, $NaNO_2$ 882
- strontium fluoride, SrF_2 832
- strontium titanate, $SrTiO_3$ 832
- sucrose 888
- sylvine 830
- sylvite 830
- TAS 850
- tellurium dioxide, TeO_2 854
- tellurium, Te 846
- tetragonal, $4/m$ (C_{4h}) 854
- tetragonal, $4/mmm$ (D_{4h}) 852
- tetragonal, 422 (D_4) 854
- tetragonal, $4mm$ (C_{4v}) 864
- tetragonal, $\bar{4}2m$ (D_{2d}) 856
- TGS 888
- thallium arsenic selenide, Tl_3AsSe_3 850
- titanium dioxide, TiO_2 852
- tourmaline, $(Na,Ca)(Mg,Fe)_3B_3Al_6Si_6(O,OH,F)_{31}$ 850
- triglycine sulfate, $(CH_2NH_2COOH)_3 \cdot H_2SO_4$ 888
- trigonal, 32 (D_3) 844
- trigonal, $3m$ (C_{3v}) 848
- trigonal, $\bar{3}m$ (D_{3d}) 844
- urea, $(NH_2)_2CO$ 862
- wurtzite 842
- YAG 832
- YAP 866
- YLF 854
- yttrium aluminate, $YAlO_3$ 866
- yttrium aluminium garnet, $Y_3Al_5O_{12}$ 832
- yttrium lithium fluoride, $YLiF_4$ 854
- yttrium vanadate, YVO_4 852
- YVO 852
- zinc blende 842
- zinc germanium diphosphide, $ZnGeP_2$ 862
- zinc oxide, ZnO 840
- zinc selenide, ZnSe 836
- zinc telluride, ZnTe 836
- zincite 840
difference frequency generation (DFG) 825
differential conductance as function of voltage 1043
diffraction method 39
diffusion coefficient 946–949, 956–960, 970
diffusion-controlled growth
- nanostructured materials 1064
dimensionality
- nanostructured materials 1033
dimensions
- physical quantities 4, 13
dimer bond length 995
dioxide
- zirconium 448
dipole glass 906
dipole moment 945, 947–953, 955–963, 968–971
direct gap
- group IV semiconductors and IV–IV compounds 592
direct piezoelectric effect 824
director 943
discotic liquid crystal 941, 942, 972
- physical properties 972
disk-like molecule 942
disordered materials 38
dispersion
- glasses 543
dispersion curves
- electronic structure of surfaces 999
dispersion hardening 329
displacement of atoms

- surface phonons 1012
displacive disorder 38
displacive ferroelectrics 907
display lifetime 944
dissipation factor 826, 828
dissipative dispersion 907
dissociation energy of molecule
- elements 46
DOBAMBC (liquid crystal) 934
domain pattern
- perpendicular magnetization 1062
domain wall
- ferroelectric 907
- ferromagnetic 907
donor surface state 1022
doping
- chemical 576
DOS (density of states) 330, 1049
- nanostructured materials 1034
drain 1024
Drude-model metal 1045
drug delivery 944
DSP family 931
DTA (differential thermal analysis) 39
dubnium Db
- elements 105
ductile iron 269
Duran
- glasses 527, 531, 537
Dy dysprosium 142
dye dopant 944
dynamic viscosity 945–948, 955–959, 961, 962, 967, 969–971, 973
dysprosium Dy
- elements 142

E

E7
- liquid crystals 975
ECS 1011
EDX (energy-dispersive analysis of X-rays) 39
EELS 1013
effect of solute elements conductivity of Cu 297
effective masses
- aluminium compounds 617
- boron compounds 607
- cadmium compounds 681
- group IV semiconductors and IV–IV compounds 593
- indium compounds 646
- mercury compounds 689

– oxides of Ca, Sr, and Ba 661
– zinc compounds 670
effective masses m_n and m_p
– gallium compounds 629
einsteinium Es
– elements 151
elastic compliance 828
elastic compliance tensor 934
– elements 47
elastic constant c_{ik} 823
– aluminium compounds 611
– beryllium compounds 652
– cadmium compounds 676
– gallium compounds 621
– group IV semiconductors and IV–IV compounds 580
– indium compounds 638
– magnesium compounds 656
– mercury compounds 687
– oxides of Ca, Sr, and Ba 660
– zinc compounds 665
elastic modulus 478, see elastic constant 580
– elements 46, 47
– polymers 478
elastic stiffness 828
– elements 47
elastic tensor 827
elastooptic coefficient 825
elastooptic constant 828
elastooptic tensor 827
electric strength 478
– polymers 478
electrical conductivity
– aluminium compounds 618
– boron compounds 608
– elements 46
– group IV semiconductors and IV–IV compounds 595
electrical conductivity see also electrical resistivity 659
electrical conductivity σ
– indium compounds 647
– magnesium compounds 659
– mercury compounds 690
– oxides of Ca, Sr, and Ba 663
electrical resistivity
– boron compounds 609
– elements 48
– gallium compounds 630
electrical steel 766
electroforming 288
electromagnetic and optical properties
– group IV semiconductors and IV–IV compounds 601

– III–V compounds 610, 619, 635, 650
– II–VI compounds 655, 660, 664, 672, 683, 691
electromagnetic concentration effect 1046
electromagnetic confinement
– nanostructured materials 1044
electromechanical coupling constant 912, 918
electron affinity
– elements 46
electron and hole mobilities
– aluminium compounds 618
electron density of states 1034
– cadmium compounds 680
electron diffraction 41
electron effective mass m_n 593, see effective mass 661
electron g-factor g_c
– gallium compounds 629
– indium compounds 646
electron microscope image
– magnetic tunneling junction 1057
electron microscopy 1049
electron mobility μ_n 663, see mobility μ 682
– elements 48
– gallium compounds 631
– indium compounds 648
– mercury compounds 690
– oxides of Ca, Sr, and Ba 663
electron transport phenomena
– nanostructured materials 1042
electron tunneling
– phonon-assisted 1043
electronegativity
– elements 46
electronic band gap
– elements 48
electronic conductivity σ
– zinc compounds 671
electronic configuration
– elements 46
electronic dispersion curves 1000
electronic ground state
– elements 46
electronic properties
– group IV semiconductors and IV–IV compounds 589–594
– III–V compounds 606, 614, 626, 643
– II–VI compounds 653, 657, 661, 668, 678, 688

electronic structure
– solid surfaces 996
electronic transport, general description
– aluminium compounds 617
– beryllium compounds 655
– boron compounds 608
– cadmium compounds 682
– gallium compounds 629
– group IV semiconductors and IV–IV compounds 595
– indium compounds 647
– magnesium compounds 659
– mercury compounds 689
– oxides of Ca, Sr, and Ba 663
– zinc compounds 670
electronic work function
– elements 48
electronic, electromagnetic, and optical properties
– elements 46
electron–phonon coupling 1040, 1041
electrooptic coefficients 826
electrooptic modulators
– nanostructured materials 1040
electrooptic tensor 827
electro-optical constants
– zinc compounds 675
electro-optical effect 944
electrostrictive constant 930
elemental semiconductor 1003
elements 45
– allotropic modifications 48
– atomic properties 46
– electromagnetic properties 48
– electronic properties 48
– high-pressure modifications 48
– ionic properties 46
– macroscopic properties 46
– materials data 46
– molecular properties 46
– optical properties 48
– ordered according to the Periodic table 52
– ordered by their atomic number 51
– ordered by their chemical symbol 50
– ordered by their name 49
elinvar alloys 786
– antiferromagnetic 789
elinvar-type alloys
– nonmagnetic 792
energy bands see band structure 590

energy diagram for a MIM tunnel
    junction   1053
energy dispersion curve   1000
energy equivalents   24
energy equivalents in different units
    24
energy exchange time τ_{e-ph} 1041
energy gap
– aluminium compounds   616
– beryllium compounds   654
– cadmium compounds   681
– gallium compounds   628
– group IV semiconductors   592,
    593
– indium compounds   644
– IV–IV compound semiconductors
    592, 593
– magnesium compounds   659
– mercury compounds   688
– oxides of Ca, Sr, and Ba   661
– zinc compounds   669
energy gaps
– boron compounds   607
energy shifts in the luminescence
    peaks   1037
energy-storage cell   1043
engineering critical current density
    741
enthalpies of phase transitions
    946–973
enthalpy change
– elements   47
enthalpy of combustion   477
– polymers   477
enthalpy of fusion   477
– polymers   477
entropy of fusion   477
– polymers   477
Er erbium   142
erbium Er
– elements   142
Es einsteinium   151
Eu europium   142
europium Eu
– elements   142
EXAFS (extended X-ray atomic
    fine-structure analysis)   39
excess carrier density   1023
excitation energy
– nanostructured materials
    1038
exciton binding energy
– gallium compounds   628
– zinc compounds   669
exciton Bohr radii
– semiconductors   1037

exciton energy
– group IV semiconductors and
    IV–IV compounds   593
exciton peak energy
– indium compounds   645
exciton Rydberg series   1038
excitons
– nanostructured materials
    1036
external forces   46
external-field dependence   46
extinction coefficient k
– gallium compounds   635
– zinc compounds   674

F

F fluorine   118
F2
– optical glasses   551
F5
– glasses   537
facets
– crystallography   27
Fahrenheit   48
families of ferroelectrics   909
fast displays   903
fcc positions
– surface diagrams   981
Fe iron   131
$Fe_{14}Nd_2B$
– commercial magnets   804
– magnetic materials   803
Fe–C(-X)
– carbide phases   224
Fe–Cr alloy   226
Fe–Mn alloys   226
Fe–Ni alloys   225
Fermi energy   998
Fermi function   1022
Fermi level pinning   1025
Fermi surface
– nanostructured materials   1055
Fermi surface shift in an electric field
    1050
Fermi surfaces
– nanostructured materials   1049
Fermi wavelength
– nanostructured materials   1035
fermium Fm
– elements   151
ferrielectric triple hysteresis loop
    936
ferrielectricity   927
ferrite
– applications   812

– hard magnetic   813
– MnZn   812
– NiZn   813
– soft magnetic   811
ferroelectric ceramics   906
ferroelectric hysteresis loop
    904
ferroelectric liquid crystal   906, 911,
    945, 967
– physical properties   967
ferroelectric mixtures
– liquid crystals   975
– physical properties   975
ferroelectric phase transition   906,
    933
ferroelectric polymers   906
ferroelectric transducer   906
ferroelectrics
– classification   906
– definition   903
– dielectric properties   903
– displacive type   906
– elastic properties   903
– families   909, 911
– general properties   906
– indirect type   906
– inorganic crystals   903
– inorganic crystals other than oxides
    922
– inorganic crystals oxides   912
– liquid crystals   903, 930
– order–disorder type   906
– organic crystals   903, 930
– phase transitions   903
– piezoelectric properties   903
– polymers   903, 930
– pyroelectric properties   903
– symbols and units   912
ferromagnetic surface   1008
Fe–Co–Cr   795
Fe–Co–V   797
Fe–Nd–B   800
– phase relations   800
Fe–Ni–Al–Co   798
Fe–Si alloys
– rapidly solidified   768
Fibonacci sequence   35
field-effect mobility   1024
– solid surfaces   1024
first Brillouin zone   see Brillouin
    zone 657
flat glasses   528
flat-band condition   1020
fluorinated three-ring LC   944
fluorine F
– elements   118

Subject Index

fluoropolymers 480, 496
– poly(ethylene-co-chlorotrifluoroethylene) (ECTFE) 496, 497
– poly(ethylene-co-tetrafluoroethylene) (ETFE) 496, 497
– poly(tetrafluoroethylene-co-hexafluoropropylene) (FEP) 496, 497
– polychlorotrifluoroethylene (PCTFE) 496, 497
– polytetrafluoroethylene (PTFE) 496, 497
flux flow (FF) 718
Fm fermium 151
formation curve
– glasses 524
formulas
– crystallographic 986
Fotoceram
– glasses 559
Fotoform
– glasses 559
Foturan
– glasses 559
Fourier map 926
four-ring system 964
Fr francium 59
francium Fr
– elements 59
Frantz–Keldysh effect 1040
free dielectric constant 907
frequency conversion 825
Fresnel reflectivity
– glasses 549
Friedel–Creagh–Kmetz rule 943
fundamental constants 3
– 2002 adjustment 4
– alpha particle 9
– atomic physics and particle physics 7
– CODATA recommended values 4
– electromagnetic constants 6
– electron 7
– meaning 4
– most frequently used 4
– neutron 8
– proton 8
– recommended values 3, 4
– thermodynamic constants 6
– units of measurement 3
– universal constants 5
– what are the fundamental constants? 3

fused silica
– glasses 534, 537

G

Ga gallium 78
GaAs positions
– surface diagrams 983
gadolinium Gd
– elements 142
gallium antimonide
– crystal structure, mechanical and thermal properties 621
– electromagnetic and optical properties 635
– electronic properties 626
– transport properties 631
gallium arsenide
– crystal structure, mechanical and thermal properties 621
– electromagnetic and optical properties 635
– electronic properties 626
– transport properties 631
gallium compounds
– crystal structure, mechanical and thermal properties 621
– electromagnetic and optical properties 635
– electronic properties 626
– mechanical and thermal properties 621
– thermal conductivity 634
– thermal properties 621
– transport properties 629
gallium Ga
– elements 78
gallium nitride
– crystal structure, mechanical and thermal properties 621
– electromagnetic and optical properties 635
– electronic properties 626
– transport properties 631
gallium phosphide
– crystal structure, mechanical and thermal properties 621
– electromagnetic and optical properties 635
– electronic properties 626
– transport properties 631
gamma titanium aluminides 213
gas permeation 478
– polymers 478
Gd gadolinium 142
Ge germanium 88

germanium
– band structure 590
– crystal structure, mechanical and thermal properties 578–588
– electromagnetic and optical properties 601
– electronic properties 589–594
– transport properties 598
germanium Ge
– elements 88
g-factor
– cadmium compounds 681
g-factor, conduction electrons
– group IV semiconductors and IV–IV compounds 594
glass designation 544
glass formers 527
glass matrix 1040
glass number 8nnn 534, 537
glass number nnnn
– sealing glasses 563
glass structure
– sodium silicate glasses 524
glass temperature
– glasses 524
glass transition temperature 477
– polymers 477
glass-ceramics 525, 526, 558
– density 558
– elastic properties 558
– manufacturing process 558
glasses 523
– Abbe value 547
– abbreviating glass code 543
– acid attack 532
– acid classes 533
– alkali attack 532
– alkali classes 533
– alkali–alkaline-earth silicate 530
– alkali–lead silicate 530
– alkaline-earth aluminosilicate 530
– amorphous metals 523
– armor plate glasses 529
– automotive applications 529
– band pass filters 566
– Borofloat 528, 529
– borosilicate 529, 530
– borosilicate glasses 529
– brittleness 536
– chemical constants 553
– chemical properties 549
– chemical resistance 549
– chemical stability 530, 531
– chemical vapor deposition 523
– color code 554
– composition 527

– compound glasses 529
– container glasses 528, 529
– crack effects 536
– density 528
– dielectric properties 538
– Duran® 531
– elasticity 536
– electrical properties 537
– engineering material 523
– fire protecting glasses 529
– flat 528
– fracture toughness 537
– frozen-in melt 533
– halide glasses 568
– hydrolytic classes 533
– infrared transmitting glasses 568
– infrared-transmitting 571
– inhomogeneous 525
– internal transmission 554
– linear thermal expansion 536, 556
– long pass filters 566
– major groups 526
– manufacturers, preferred optical glasses 545
– melting range 533
– mixtures of oxide compounds 524
– neutral density filters 566
– optical 551
– optical characterization 547
– optical glasses 543
– optical properties 539, 543
– oxide glasses 568
– passivation glasses 562
– physical constants 553
– plate glasses 528
– properties 526
– quasi-solid melt 533
– refractive index 539
– Schott filter glasses 569
– sealing glasses 559
– short pass filters 566
– silicate based 526
– soda–lime type 528
– solder glasses 562
– strength 534
– stress behavior 535
– stress rate 535
– stress-induced birefringence 539
– supercooled melt 533
– surface (cleaning and etching) 533
– surface modification 532
– surface resistivity 538
– technical 530
– tensile strength 536

– thermal conductivity 556
– thermal strength 537
– transmittance 539
– viscosity 534
– vitreous silica 556
– volume resistivity 537
– wear-induced surface defects 535
glasses, colored
– nomenclature 566
– optical filter 566
glasses, sealing
– ceramic 562
– principal applications 561
– recommended material combinations 560
– special properties 561
glasses, solder and passivation
– composite 563
– properties 564
glassy state
– crystallography 39
GMO family
– ferroelectrics 920
GMR (giant magnetoresistance) 1049, 1050
– mechanism 1051
– thickness dependence 1051
GMR ratio 1050
gold
– alloys 347
– applications 347
– chemical properties 361
– electrical properties 356
– electrical resistivity 356
– intermetallic compounds 350
– magnetic properties 358
– mechanical properties 352
– optical properties 359
– phase diagrams 347
– production 347
– special alloys 361
– thermal properties 359
– thermochemical data 347
– thermoelectric properties 358
gold Au
– elements 65
golden mean 35
granular materials 1043
gray
– SI unit of absorbed dose 19
gray iron 269
gray tin
– band structure 590
– crystal structure, mechanical and thermal properties 578–588

– electromagnetic and optical properties 601
– electronic properties 589–594
– transport properties 599
Griffith flaw
– glasses 534
Group IV semiconductors 576, 578–603
– electron mobility 597
– hole mobility 597
Group IV semiconductors and IV–IV compounds
– crystal structure 578
– electromagnetic and optical properties 601
– electronic properties 589–594
– mechanical properties 578
– thermal properties 578
– transport properties 595–601
groups of elements (Periodic table) 45

H

H hydrogen 54
hafnium Hf
– elements 94
Haigh Push-Pull test 408
Hall coefficient
– elements 48
– group IV semiconductors and IV–IV compounds 595
Hall mobility
– group IV semiconductors and IV–IV compounds 596, 597
Hall mobility see also mobility μ 596
halogen-substituted benzene 946
hard disk drive 1060
– limits 1060
– technology 1060
hard ferrites
– magnetic properties 814
hard magnetic alloys 794
hardenability 237
hardmetals 277
hassium Hs
– elements 131
Hatfield steel 226
HATOF 1013
HCl family 924
He helium 54
heat capacities c_p, c_V
– boron compounds 605
– group IV semiconductors 584

– IV–IV compound semiconductors 584
heat capacity   477, 956–960, 970
– cadmium compounds   678
– gallium compounds   623
– indium compounds   640
– mercury compounds   687
– oxides of Ca, Sr, and Ba   661
– polymers   477
– zinc compounds   667
heat-resistant steels   258
HEIS (high-energy ion scattering) 989, 1013
helical structure   942
helium He
– elements   54
Hermann–Mauguin symbols   30
hertz
– SI unit of frequency   19
heterocycles   946
hexagonal BaTiO$_3$ 917
Hf hafnium   94
Hg mercury   73
high copper alloy   297
high-T_c superconductors
– lower critical   716
– upper critical   716
high-frequency dielectric constant ε
  see also dielectric constant ε 601
high-Ni alloys   281
high-pressure die casting (HPDC) 168
high-pressure modifications   48
high-strength low-alloy   240
hip implants   277
HMF (half-metallic ferromagnet) 1055
Ho holmium   142
hole effective mass m_p 594, see effective mass 661
hole mobility μ_p see mobility μ 682
– elements   48
– gallium compounds   631
– indium compounds   648
hollow-ware
– glasses   528
holmium Ho
– elements   142
holohedry
– crystallography   30
homeotropic alignment   943
Hooke's law   47, 823
hopping mechanism
– nanostructured materials   1043
host–guest effect   944
hot forming

– glasses   524
Hoya code
– glasses   544
HPDC (high-pressure die casting) 168
HRTEM (high-resolution transition electron microscopy)   39
Hs hassium   131
Hume-Rothery phase   333, 350
Hume-Rothery phases   296
hydrogen H
– elements   54
hydrostatic pressure   926
hyper-Raman scattering   915

I

I iodine   118
IBA   991, 1024
ICSU (International Council of the Scientific Unions)   4
ideal surface   979
III–V compound semiconductors 604
II–VI semiconductor compounds 652
image potential   996
image state   996, 997, 999
– effective mass   999
impact strength   478
– polymers   478
impurity elements   206
impurity scattering
– group IV semiconductors and IV–IV compounds   599
In indium   78
incommensurate phase   930
incommensurate phases   906
incommensurate reconstruction   986
index of refraction
– complex   674
indirect gap
– group IV semiconductors and IV–IV compounds   589
indium antimonide
– crystal structure, mechanical and thermal properties   638
– electromagnetic and optical properties   650
– electronic properties   643
– transport properties   647
indium arsenide
– crystal structure, mechanical and thermal properties   638
– electromagnetic and optical properties   650

– electronic properties   643
– transport properties   647
indium compounds
– crystal structure, mechanical and thermal properties   638
– electromagnetic and optical properties   650
– electronic properties   643
– mechanical and thermal properties 638
– optical properties   650
– thermal properties   638
– transport properties   647
indium In
– elements   78
indium nitride
– crystal structure, mechanical and thermal properties   638
– electromagnetic and optical properties   650
– electronic properties   643
– transport properties   647
indium phosphide
– crystal structure, mechanical and thermal properties   638
– electromagnetic and optical properties   650
– electronic properties   643
– transport properties   647
induced phase transition   904
inelastic neutron scattering   906
infrared-transmitting glasses   568, 571
inorganic ferroelectrics   903
inorganic ferroelectrics other than oxides   922
inosilicates   433
insulator
– surface phonon   1017
intercritical annealing   240
interface state density
– solid surface   1026
interlayer distance
– crystallographic formulas   986
International Annealed Copper Standard (IACS)   296
international system of units   11
international tables for crystallography   31
International Union of Pure and Applied Chemistry (IUPAC)   15
International Union of Pure and Applied Physics (IUPAP)   14
internuclear distance
– elements   46
intrinsic carrier concentration

- group IV semiconductors and IV–IV compounds 595, 596
intrinsic charge carrier concentration
- elements 48
intrinsic Debye length 1021
intrinsic Fermi level 1020
invar alloys
- Fe–Ni-based 782
- Fe–Pd base 785
- Fe–Pt-based 784
Invar effect 385
inversion center 30
inversion layer 1020
- solid surfaces 1024
inversion layer channel 1024
iodine I
- elements 118
ionic radius
- elements 46, 49
ionization energy
- elements 46
Ir iridium 135
iridium 393
- alloys 393
- applications 393
- chemical properties 398
- diffusion 398
- electrical properties 397
- lattice parameter 394
- magnetic properties 397
- mechanical properties 395
- optical properties 398
- phase diagram 393
- production 393
- thermal properties 398
- thermoelectrical properties 397
iridium Ir
- elements 135
iron and steels 221
iron Fe
- elements 131
iron miscibility gap 226
iron phase diagram 226
iron-carbon alloys 222
iron-cobalt alloys 772
iron–silicon alloys 763
Ising model 906
ISO (International Organization for Standardization) 12
isothermal transformation (IT) diagram 238
isotropic
- dielectrics 828
isotropic liquid 943
IV–IV compound semiconductors 576, 578–603

- electron mobility 597
- hole mobility 597

J

JDOS 1006
jellium 997
jellium model 998
- work functions 999
jewellery 330
joint density of states 1005
jominy apparatus 238
Josephson vortices 717
joule
- SI unit of energy 19

K

K 50
- glasses 537
K potassium 59
K10
- optical glasses 551
K7
- optical glasses 551
katal
- SI unit of catalytic activity 19
KDP family 925
kelvin 48
- SI base unit 14
Kerr effect 825
- optical 826
Kerr ellipticity 1011
KH_2PO_4 family 925
kilogram
- SI base unit 14
kinematic viscosity 945, 947, 950, 951, 955, 962, 964–967, 969, 975
Kleinman symmetry conditions 825
$KNbO_3$ 913
knee joint replacements 277
KNO_3 family 925
Knoop hardness
- optical glasses 550
Kr krypton 128
KRIPES 997
Kroll process 206
krypton Kr
- elements 128
$KTaO_3$ 913

L

LA
- phonon spectra 915
La lanthanum 84

Lamb theory of elastic vibrations 1040
lamellar (flake) graphite (FG) 268
langbeinite-type family 911, 930
lanthanum La
- elements 84
LASF35
- optical glasses 551
LAT family 932
lattice concept
- crystallography 28
lattice constants see lattice parameters 656
lattice dynamics 1012
lattice parameter 980
- aluminium compounds 610
- beryllium compounds 652
- boron compounds 604
- cadmium compounds 676
- gallium compounds 621
- group IV semiconductors and IV–IV compounds 579
- indium compounds 638
- magnesium compounds 656
- mercury compounds 686
- oxides of Ca, Sr, and Ba 660
- zinc compounds 665
lattice scattering
- group IV semiconductors and IV–IV compounds 598
lattice vibration 906
lattices
- planes and directions 28
Laue images 41
lawrencium Lr
- elements 151
layer-structure family 909
LB (Langmuir–Blodgett) film 944
LC display 944
LC materials (LCMs) 944
LC–surface interaction 944
lead 407
- antimony 412
- arsenic alloys 421
- battery grid alloys 413
- bearing alloys 415
- bismuth 419
- cable sheathing alloys 421
- calcium–tin 417
- calcium–tin, battery grid 418
- coper alloys 421
- corrosion 408
- corrosion classification 411
- fusible alloys 420
- gamma-ray mass-absorption 412
- grades 407

– internal friction 408
– low-melting alloys 419
– mechanical properties 408
– quaternary eutectic alloy 420
– recrystallization 409
– silver alloys 420
– solder alloys 415
– solders 416
– tellurium alloys 421
– ternary alloys 413
– tin alloy 415
lead glasses 527
lead Pb
– elements 88
lead–antimony
– phase diagram 413
LEED (low-energy electron diffraction) 987, 1013
LEIS (low-energy ion scattering) 989, 1013
LF5
– optical glasses 551
Li lithium 59
$Li_2Ge_7O_{15}$ family 920
$LiBaO_3$ 919
light transmittance
– glasses 539
light-emitting diode 1043
$LiNbO_3$ family 909, 919
linear thermal expansion
– optical glasses 556
linear thermal expansion coefficient α
– aluminium compounds 611
– beryllium compounds 653
– boron compounds 605
– cadmium compounds 677
– gallium compounds 623
– group IV semiconductors 582, 583
– indium compounds 639
– IV–IV compound semiconductors 582, 583
– magnesium compounds 656
– mercury compounds 687
– oxides of Ca, Sr, and Ba 660
– zinc compounds 666
$LiNH_4C_4H_4O_8$ family 911
liquid crystal
– anisotropy 943
– refractive index 943
– rod-like 943
liquid crystal family 911, 934
liquid crystal ferroelectrics 903
liquid crystal material (LCM)
– degradation 944
liquid crystal salts 973

– physical properties 973
liquid crystal two-ring systems with bridges
– physical properties 955
liquid crystal two-ring systems without bridges
– physical properties 947
liquid crystalline acids
– physical properties 946
liquid crystalline compound 943
– mesogenic group 943
– side group 943
– terminal group 943
liquid crystalline mixtures
– physical properties 975
liquid crystals
– ferroelectric properties 905
liquid crystals (LCs) 941
list of described physical properties 822
lithium Li
– elements 59
lithium niobate 919
lithography
– nanostructured materials 1064
Lithosil^{TM} 551
LLF1
– optical glasses 551
local-field effect 1045
long pass filters
– glasses 566
longitudinal acoustic branch 915
long-range order 39, 927
– glasses 524
long-range order of molecules 943
losses
– dynamic eddy current 763
– hysteresis 763
low dielectric loss glasses 527
low-dimensional system
– nanostructured materials 1034
low-frequency dielectric constant 917
low-temperature annealing 298
Lr lawrencium 151
LSMO ($La_{0.7}Sr_{0.3}MnO_3$) 1056
Lu lutetium 142
lumen
– non-SI unit in photometry 15
luminescence
– nanostructured materials 1036
luminous flux
– photometry 15
luminous intensity I_v
– photometry 15
lutetium Lu

– elements 142
lyotropic liquid crystals 942

M

M. Faraday 1032
Macor
– glasses 559
magnesium alloys 163
– corrosion behavior 169
– heat treatments 169
– joining 169
– mechanical properties 168
– nominal composition 165
– solubility data 163
– tensile properties 167
– tensile property 166
magnesium compounds
– crystal structure, mechanical and thermal properties 655
– electromagnetic and optical properties 660
– electronic properties 657
– mechanical and thermal properties 655
– optical properties 660
– thermal properties 655
– transport properties 659
magnesium Mg
– casting practices 168
– elements 68
– magnesium alloys 162
– melting practices 168
magnesium oxide 444
– applications 444
– crystal structure, mechanical and thermal properties 655
– electrical properties 444
– electromagnetic and optical properties 660
– electronic properties 657
– mechanical properties 444
– thermal properties 444
– transport properties 659
magnesium selenide
– crystal structure, mechanical and thermal properties 655
– electromagnetic and optical properties 660
– electronic properties 657
– transport properties 659
magnesium silicate
– electrical properties 435, 436
– mechanical properties 435, 436
– thermal properties 435, 436
magnesium sulfide

- crystal structure, mechanical and thermal properties 655
- electromagnetic and optical properties 660
- electronic properties 657
- transport properties 659

magnesium telluride
- crystal structure, mechanical and thermal properties 655
- electromagnetic and optical properties 660
- electronic properties 657
- transport properties 659

magnet
- Mn−Al−C 811

magnetic domain 1058
magnetic dots
- nanostructured materials 1049
magnetic dots, arrays of 1061
magnetic field constant
- fundamental constant 14
magnetic layers 1048, 1050
- spin valve 1052
magnetic materials 755
- Co_5Sm based 806
- hard 794
- permanent 794
magnetic nanostructures 1031, 1048, 1050
- information storage 1048
- read heads 1048
- sensors 1048
magnetic oxides 811
magnetic periodic structures 33
magnetic reading head 1053, 1058
magnetic recording
- perpendicular discontinuous media 1060
magnetic sensors 1048, 1058
magnetic surface 1008
magnetic susceptibility 948–950, 956, 957
- elements 46, 48
magnetic tunnel junction 1054
- manganite-based 1057
magnetic tunneling junctions (MTJ) 1056
magnetization
- elements 48
magnetocrystalline anisotropy 756
magnetoelectronic devices 1058
magnetoresistance effect 1050
magnetoresistance of Fe/Cr multilayers 1051
magnetostriction 757

magnets
- 17/2 type 807
- 5/1 type 806
- $TM_{17}Sm_2$ 808
majority carriers 1023
malleable irons 270
manganese Mn
- elements 124
manipulating the dot magnetization 1062
manufacturing
- Fe−Nd−B magnets 801
manufacturing process
- glasses 525
martensite 223
martensitic transformation 225, 226
mass magnetic susceptibility
- elements 48
mass susceptibility
- elements 48
mass-production glasses 526
materials
- semiconductors 576
materials data
- elements 46
material-specific parameters 46
matrix composites 170
Maxwell–Garnett model 1045
Maya blue color 1032
MBBA
- liquid crystals 955
MBE (molecular-beam epitaxy) 991, 1032, 1049, 1063
- 0-D structures 1064
- 1-D structures 1064
- 2-D structures 1063
Md mendelevium 151
mechanical and thermal properties
- III–V compounds 621, 638
- II–VI compounds 652, 655, 660, 665, 676, 686
mechanical properties
- elements 46, 47
- group IV semiconductors 578–588
- III–V compounds 610
- III–V semiconductors 604
- IV–IV compound semiconductors 578–588
- optical glasses 550
- technical glasses 533
MEIS (medium-energy ion scattering) 989, 1013
meitnerium Mt
- elements 135
melt viscosity 478

- polymers 478
melting point T_m
- aluminium compounds 611
- beryllium compounds 653
- boron compounds 605
- cadmium compounds 677
- gallium compounds 622
- group IV semiconductors and IV–IV compounds 580
- indium compounds 639
- magnesium compounds 656
- mercury compounds 686
- oxides of Ca, Sr, and Ba 660
- zinc compounds 666
melting temperature 477
- elements 47, 48
- polymers 477
memory devices 903
mendelevium Md
- elements 151
mercury compounds
- crystal structure, mechanical and thermal properties 686
- electromagnetic and optical properties 691
- electronic properties 688
- mechanical and thermal properties 686
- thermal properties 686
- transport properties 689
mercury Hg
- elements 73
mercury oxide
- crystal structure, mechanical and thermal properties 686
- electromagnetic and optical properties 691
- electronic properties 688
- transport properties 689
mercury selenide
- crystal structure, mechanical and thermal properties 686
- electromagnetic and optical properties 691
- electronic properties 688
- transport properties 689
mercury sulfide
- crystal structure, mechanical and thermal properties 686
- electromagnetic and optical properties 691
- electronic properties 688
- transport properties 689
mercury telluride
- crystal structure, mechanical and thermal properties 686

– electromagnetic and optical properties 691
– electronic properties 688
– transport properties 689
mesogen 943
mesogenic group 943
mesophase 941
mesoscopic material
– conductivity 1044
– nanoparticle doped 1044
– waveguide applications 1045
mesoscopic materials 1031, 1033
– manufacturing 1033
mesoscopic system
– quantum size effect 1035
– thermodynamic stability 1035
metal
– nanoparticle 1040
– resonance state 999
– surface 987
– surface core level shifts (SCLS) 998
– surface Debye temperature 1013
– surface phonon 1012
– surface state 999
– work function 997
metal surface 987
– jellium model 998
metals 997
– vertical relaxation 989
meter
– SI base unit 13
metrologica, international journal 12
MFM image of a written line on an array of dots 1063
MFM image of arrays of dots 1063
MFM image of domain pattern 1062
– sidewalls 1062
Mg magnesium 68
MHPOBC (liquid crystal) 935, 967
microphase separation 941
Mie theory 1045
Miller delta 825
Miller indices 28
M–I–M (metal–insulator–metal) heterostructure 1053
minority carriers 1023
missing-row reconstruction 987
Mn manganese 124
Mn–Al–C 810
– phase relations 810
Mo molybdenum 114
Mo-based alloys 317
mobility μ

– aluminium compounds 618
– cadmium compounds 682
– group IV semiconductors and IV–IV compounds 597–599
– indium compounds 648
MOCVD (metal-organic chemical vapor deposition) 1064
modulated crystal structure 924
modulated structures
– crystallography 34
moduli see elastic constant 580
Mohs hardness 822, 826
– elements 47
molar enthalpy of sublimation
– elements 47
molar entropy
– elements 47
molar heat capacity
– elements 47
molar magnetic susceptibility
– elements 48
molar mass 945–972
– glasses 527
molar susceptibility
– elements 48
molar volume
– elements 47
mole
– definition 14
– SI base unit 14
mole fraction
– glasses 527
molecular architecture 477
molybdenum Mo
– elements 114
momentum-conservation rule 1036
monocrystalline material 47
monolithic alloys 170
monophilic liquid crystal 942
MOS devices 979
MOS field-effect transistor 1020
MOSFET
– electron and hole mobility 1025
– equilibrium condition 1024
– schematic drawing 1025
Mott–Wannier exciton 1037
MQW (multiple quantum well) 1038
MRAM (magnetic random access memories) 1058
MRAM cell
– schematic diagram 1059
Mt meitnerium 135
MTJ (magnetic tunnel junction) 1053, 1054
MTJ sensor

– operation principle 1059
multi-component alloys 219
multiphase (MP) alloys 276
multiple hysteresis loops 932
multiple quantum wells 1042
MVA-TFT (multidomain vertical-alignment thin-film transistor) 944

N

N nitrogen 98
N16B
– glasses 537
– sealing glasses 563
N-4
– liquid crystals 959
Na sodium 59
$NaNO_2$ family 924
nanocrystal 1038
nanoimprint lithography 1061
nanoimprinted single domain dots
– images 1061
nanoimprinting 1061
nanolithography 1031
nanomaterial 1032
nanometric multilayers 1049
nanoparticle 1031
– doped material 1045
– local-field 1045
nanopatterning 1061
nanophase materials 1032
nanoporous materials 1032, 1043
nanoscience 1032
nanostructured material 1031, 1032
– classification scheme 1034
– conductance 1043
– definition 1032
– electrical conductivity 1043
– manufacturing 1063
– preparation 1031, 1035
– zeolites 1065
nanostructures
– magnetic 1048
nanotechnology 1032
Nb niobium 105
N-BAF10
– optical glasses 551
N-BAF52
– optical glasses 551
N-BAK4
– optical glasses 551
N-BALF4
– optical glasses 551
N-BASF64
– optical glasses 551

Nb-based alloys 318
N-BK7
– glasses 548
– optical glasses 551
Nd neodynium 142
Nd–Fe–B
– physical properties 802
Ne neon 128
near field microscopy 1049
nematic mixtures
– liquid crystals 975
– physical properties 975
nematic phase
– liquid crystals 941
nematic–isotropic transition 945
nematic-phase director 943
Neoceram
– glasses 559
neodynium Nd
– elements 142
neon Ne
– elements 128
Neoparies
– glasses 559
neptunium Np
– elements 151
nesosilicate 433
neutral density filters
– glasses 566
neutron diffraction 41, 926
neutron scattering 915
neutron spectrometer scan 916
new rheocast process (NRC) 170
newton meter
– SI unit of moment of force 19
N-FK51
– glasses 548
– optical glasses 551
N-FK56
– optical glasses 551
$(NH_4)_2SO_4$ family 927
$(NH_4)_3H(SO_4)_2$ family 928
$(NH_4)HSO_4$ family 928
$(NH_4)LiSO_4$ family 928
Ni nickel 139
Ni superalloys 294
nickel
– alloys 279
– application 279
– carbides 285
– low-alloy 279
– mechanical properties 280
– plating 288
nickel Ni
– elements 139
nickel-based superalloys 284

nickel–iron alloys 769
nickel-silvers 300
NIMs (National Institutes for Metrology 4
niobium Nb
– elements 105
nitride
– electrical properties 467
– mechanical properties 467
– thermal properties 467
nitrides
– physical properties 468
nitrogen N
– elements 98
N-KF9
– optical glasses 551
N-KZFS2
– optical glasses 551
N-LAF2
– optical glasses 551
N-LAK33
– optical glasses 551
N-LASF31
– optical glasses 551
No nobelium 151
nobelium No
– elements 151
noble metals 329
– Ag 329
– alloys 329
– applications 330
– Au 329
– catalysts 330
– corrosion resistance 329
– hardness 329
– Ir 329
– optical reflectivity 330
– Os 329
– Pd 329
– Pt 329
– Rh 329
– Ru 329
– vapour pressure 330
NOL (nano-oxide layer) 1053
noncrystallographic diffraction symmetries 36
nondestructive testing 941
nonlinear field-dependent properties 46
nonlinear optical coefficients
– nanostructured materials 1039
nonlinear optical device 903, 925
nonlinear optical susceptibility 920
nonlinear susceptibility tensors 827
non-oxide ferroelectrics 905
non-SI units 11, 20

normalizing 224
Np neptunium 151
N-PK51
– optical glasses 551
N-PSK57
– optical glasses 551
N-SF1
– optical glasses 551
N-SF56
– optical glasses 551
N-SF6
– glasses 548
– optical glasses 551
N-SK16
– optical glasses 551
N-SSK2
– optical glasses 551
nuclear incoherent scattering 916
nuclear reactor 218

O

O oxygen 108
occupied electron shells
– elements 46
Ohara code
– glasses 544
one-dimensional liquid 943
one-electron potential 999
opal
– nanostructured materials 1032
opals 1048
OPO (optical parametric oscillation) 825
optical constants
– gallium compounds 635
– group IV semiconductors and IV–IV compounds 601–603
– indium compounds 651
optical constants n and k
– cadmium compounds 684
optical glasses 526, 543
– thermal properties 556
optical materials
– high-frequency properties 824
optical mode frequency 916
optical parametric oscillation (OPO) 825
optical parametric oscillator 920
optical phonon scattering
– group IV semiconductors and IV–IV compounds 598
optical phonon scattering see also phonon scattering 598
optical phonon softening 913
optical properties

– group IV semiconductors and IV–IV compounds 601
– III–V compounds 610, 619, 635, 650
– II–VI compounds 655, 660, 664, 672, 683, 691
optical second-harmonic generator 903
optical transparency range 826
optical visualization 941
optoelectronic devices
– nanostructured materials 1038
order parameter 946–952, 956–960, 962, 963
order parameter, S 943
order principle for mesogenic groups 946
organic ferroelectrics 903
orientational order 941
Os osmium 131
osmium Os
– alloys 402
– applications 402
– cathodes 404
– chemical properties 406
– electrical properties 404
– elements 131
– lattice parameter 402
– magnetic properties 405
– mechanical properties 404
– phase diagrams 402
– production 402
– thermal properties 405
– thermoelectric properties 404
outer-shell orbital radius
– elements 46
over-aging 201
oxidation states
– elements 46
oxide 437
– beryllium 447
– magnesium 444
– physical properties 438
oxide ceramics
– production of 444
oxide ferroelectrics 903, 905
oxide superconductors
– low-T_c oxide 711
oxides of Ca, Sr, and Ba
– crystal structure, mechanical and thermal properties 660
– electromagnetic and optical properties 664
– electronic properties 661
– mechanical and thermal properties 660

– thermal properties 660
– transport properties 663
oxygen O
– elements 108

P

P phosphorus 98
Pa protactinium 151
PAA
– liquid crystals 958
pair distribution function
– crystallography 40
palladium Pd
– alloys 364
– applications 364
– electrical properties 370
– elements 139
– lattice parameters 366
– magnetic properties 372
– mechanical properties 368
– phase diagrams 364
– production 364
– thermoelectric properties 370
parabolic band 1000
Parkes process 330
partially stabilized zirconia (PSZ)
– electrical properties 449
– mechanical properties 449
– thermal properties 449
particle in a box model 1035
particle intensity I_p
– radiometry 15
passivation glasses 562, 564
– glasses 565
– properties 564
pattern transfer by imprinting 1061
Pauling 46
Pb lead 88
PbHPO$_4$ family 927
PbTiO$_3$ 916
PbZrO$_3$ 918
Pb–Ca–Sn
– battery grid alloys 418
PCH-7
– liquid crystals 950
Pd palladium 139
pearlite 222
Pearson symbol
– elements 47
percolation density
– nanostructured materials 1033
periodes of elements (Periodic table) 45
Periodic table
– elements 45

Periodic table of the elements 53
permanent magnets
– Co–Sm 806
perovskite-type family 909, 911, 912
perovskite-type oxide 909
phase diagram
– Fe–C 222
– Fe–Cr 226
– Fe–Mn 225
– Fe–Ni 225
– Fe–Si 227
– Ti–Al 210
– Zr–Nb 218
– Zr–O 218
phase separation
– glasses 525
phase transition temperature 943
phase-matching angle 826
phase-matching condition 825
phonon confinement
– nanostructured materials 1042
phonon density of states
– gallium compounds 623, 625
– indium compounds 642
phonon dispersion
– surface phonons 1012
phonon dispersion curve
– aluminium compounds 612
– gallium compounds 623–625
– indium compounds 641
– surface phonons 1015–1026
phonon dispersion relation
– group IV semiconductors 585–587
– IV–IV compound semiconductors 585, 588
phonon energies
– nanostructured materials 1040
– solid surfaces 1013
phonon frequencies ν
– cadmium compounds 678
– gallium compounds 624
– group IV semiconductors 585
– indium compounds 640
– IV–IV compound semiconductors 585
– magnesium compounds 657
– mercury compounds 687
– oxides of Ca, Sr, and Ba 661
– zinc compounds 667
phonon frequencies ν see also phonon wavenumbers $\tilde{\nu}$ 605
phonon instability 921
phonon mode frequency 908
phonon scattering

– group IV semiconductors and IV–IV compounds   598
phonon wavenumbers $\tilde{\nu}$ *see also* phonon frequencies ν 605
– aluminium compounds   612
– cadmium compounds   678
– gallium compounds   624
– indium compounds   640
– magnesium compounds   657
– mercury compounds   687
– zinc compounds   667
phonon wavenumbers $\tilde{\nu}$/frequencies ν
– boron compounds   605
phonon–phonon coupling   907
phosphorus P
– elements   98
photochromic glasses
– nanostructured materials   1032
photoconductive crystal   922
photoelastic effect   824
photoemission spectra for Ag quantum wells   1036
photoluminescence spectra of CdSe   1039
photometry
– intensity measurements   15
photonic band-gap materials   1048
phyllosilicate   433
physical properties
– liquid crystals   943
physical quantities   12
– base   11, 12
– data   13
– definition   12
– derived   11, 12
– general tables   4
piezoelectric
– element   918
– material   917
– strain constant   918
– strain tensor   824
– tensor   827
piezoelectricity   824, 906
piezooptic coefficient   825
planar alignment   943
planar electromechanical coupling factor   918
Planck radiator   15
plasmon excitations
– nanostructured materials   1031
plasmon oscillation
– nanostructured materials   1032
plasmon peak
– metals   1045
plasmon resonance   1045

plastic crystal   941
platinum group metals (PGM)   363
– alloys   363
platinum Pt
– alloys   376
– applications   376
– catalysis   385
– chemical properties   385
– electrical properties   381
– elements   139
– magnetical properties   384
– mechanical properties   378
– optical properties   385
– phase diagrams   376
– production   376
– thermal properties   385
– thermoelectric properties   382
plutonium Pu
– elements   151
PLZT
– ceramic material   918
Pm promethium   142
Po polonium   108
Pockels effect   825
point groups
– crystallography   30
Poisson equation   1022
Poisson number
– elements   47
Poisson's ratio   478
– optical glasses   550
– polymers   478
polarization microscopy   943
polonium Po
– elements   108
poly(4-methylpentene-1) (PMP)   488, 489
poly(ethylene-co-acrylic acid) (EAA)   486, 487
poly-(ethylene-co-norbornene)   486, 487
poly(ethylene-co-vinyl acetate) (EVA)   486, 487
poly(vinyl chloride)   492, 495
– plastisized (60/40) (PVC-P2)   492, 495
– plastisized (75/25) (PVC-P1)   492, 495
– unplastisized (PVC-U)   492, 494–496
polyacetals   480, 497
– poly(oxymethylene) (POM-H)   497–500
– poly(oxymethylene-co-ethylene) (POM-R)   497, 498, 500
polyacrylics   480, 497

– Poly(methyl methacrylate) (PMMA)   497–499
polyamides   480, 501
– polyamide 11 (PA11)   501
– polyamide 12 (PA12)   501
– polyamide 6 (PA6)   501, 502
– polyamide 610 (PA610)   501, 502
– polyamide 66 (PA66)   501, 502
polybutene-1 (PB)   488, 489
polyesters   481, 503
– poly(butylene terephthalate) (PBT)   503–505
– poly(ethylene terephthalate) (PET)   503–505
– poly(phenylene ether) (PPE)   504–506
– polycarbonate (PC)   503, 504
polyether ketones   481, 508
– poly(ether ether ketone) (PEEK)   508, 509
polyethylene   483–486
– high density (HDPE)   483–486
– linear low density (LLDPE)   484–486
– low density (LDPE)   484–486
– medium density (MDPE)   484–486
– ultra high molecular weight (UHMWPE)   484–486
polyethylene ionomer (EIM)   486, 487
polyimides   481, 508
– poly(amide imide) (PAI)   508
– poly(ether imide) (PEI)   508, 509
– polyimide (PI)   508
polyisobutylene (PIB)   488, 489
polymer
– physical properties   483
polymer blend   515
– physical properties   477, 483
– poly(acrylonitrile-co-butadiene-co-acrylester) + polycarbonate (ASA + PC)   515–517
– poly(acrylonitrile-co-butadiene-co-styrene) + polyamide (ABS + PA)   515–517
– poly(acrylonitrile-co-butadiene-co-styrene) + polycarbonate (ABS + PC)   515–517
– poly(butylene terephthalate) + poly(acrylonitrile-co-butadiene-co-acrylester) (PBT + ASA)   517, 521
– poly(butylene terephthalate) + polystyrene (PBT + PS)   515, 519, 520

– poly(ethylene terephthalate) + polystyrene (PET + PS) 515, 519, 520
– poly(phenylene ether) + polyamide 66 (PPE + PA66) 517, 521
– poly(phenylene ether) + polystyrene (PPE + PS) 520, 521
– poly(styrene-co-butadiene) (PPE + SB) 517, 520, 521
– poly(vinyl chloride) + chlorinated polyethylene (PVC + PE-C) 515, 518
– poly(vinyl chloride) + poly(acrylonitrile-co-butadiene-co-acrylester) (PVC + ASA) 515, 518
– poly(vinyl chloride) + poly(vinyl chloride-co-acrylate) (PVC + VC/A) 515, 518
– polycarbonate + liquid crystal polymer (PC + LCP) 515, 519, 520
– polycarbonate + poly(butylene terephthalate) (PC + PBT) 515, 519, 520
– polycarbonate + poly(ethylene terephthalate) (PC + PET) 515, 519, 520
– polypropylene + ethylene/propylene/diene rubber (PP + EPDM) 515, 516
– polysulfone + poly(acrylonitrile-co-butadiene-co-styrene) (PSU + ABS) 517, 521
polymer family 911, 936
polymer ferroelectrics 903
polymer matrix 1040
polymers 477
– abbreviations 482
– Charpy impact strength 478
– coefficient of expansion 478
– compressibility 478
– creep modulus 478
– crystallinity 477
– density 478
– elastic modulus 478
– electric strength 478
– enthalpy of combustion 477
– enthalpy of fusion 477
– entropy of fusion 477
– ferroelectric properties 905
– gas permeation 478
– glass transition temperature 477
– heat capacity 477
– impact strength 478
– melt viscosity 478
– melting temperature 477
– physical properties 477
– physicochemical properties 477
– Poisson's ratio 478
– refractive index 478
– relative permittivity 478
– shear modulus 478
– shear rate 478
– Shore hardness 478
– sound velocity 478
– steam permeation 478
– stress 478
– stress at 50% strain (elongation) 478
– stress at fracture 478
– stress at yield 478
– structural units 479–481
– surface resistivity 478
– thermal conductivity 478
– Vicat softening temperature 477
– viscosity 478
– volume resistivity 478
polyolefines 480, 483–486
polypropylene (PP) 488, 489
polysulfides 481, 506
– poly(phenylene sulfide) (PPS) 506, 507
polysulfones 481, 506
– poly(ether sulfone) (PES) 506, 507
– polysulfone (PSU) 506, 507
polyurethanes 481, 511
– polyurethane (PUR) 511, 512
– thermoplastic polyurethane elastomer (TPU) 511, 512
polyvinylidene fluoride
– ferroelectrics 911
porous aluminium silicates
– electrical properties 436
– mechanical properties 436
– thermal properties 436
Portland cement
– ASTM types 433
– chemical composition 432
positional order 941
potassium K
– elements 59
potential barrier 1022
powder-composite materials 277
Powder-in-Tube (PIT) 739
power-law dependence of conductivity on film thickness 1042
Pr praseodynium 142
practical superconductors
– characteristic properties 705
praseodynium Pr
– elements 142
prefixes
– decimal multiples of units 19
primitive cell
– crystal structure 28
projected band structure 996
projected bond length 995
promethium Pm
– elements 142
property tensor 46
– independent components 827
protactinium Pa
– elements 151
proton distribution 926
pseudopotential calculation 1006
Pt platinum 139
p-type diamond
– Debye length 1021
Pu plutonium 151
pulsed infrared lasers 1040
pyrochlore-type family 909
pyroelectric coefficient 917
pyroelectric measurement 931
PZT
– piezoelectric material 917

Q

quantum confinement 1036
– nanostructured materials 1042
quantum dots
– nanostructured materials 1031, 1035
quantum size effect
– nanostructured materials 1031, 1035
quantum transport
– nanostructured materials 1053
quantum well
– coupled 1043
– nanostructured materials 1031, 1035
quantum wires
– nanostructured materials 1035
quantum-well superlattices 1033
quasicrystals 34
QWIP (quantum well infrared photodetector) 1042

R

Ra radium 68
radiant intensity I_e
– radiometry 15

radiation sources and exposure
   techniques in lithography   1065
radiometric and photometric
   quantities   16
radiometry
 – intensity measurements   15
radium Ra
 – elements   68
radon Rn
 – elements   128
Raman scattering   906
Raman scattering spectroscopy
   1040
Raman spectrum   908
RAS   997
Rayleigh mode   1012
Rb rubidium
 – elements   59
Re rhenium
 – elements   124
real and imaginary parts ε_1 and ε_2 of
   the dielectric constant   *see*
   dielectric constant ε 602
 – gallium compounds   637
reconstruction model   987
 – solid surfaces   991
reconstruction of semiconductors
   991
reconstruction of surface   986
 – metals   987
recording media
 – arrays of magnetic dots   1061
reduced surface state energy   1022
reduced wave vector   1012
reduced-dimensional material
   geometries   1033
references
 – solid surfaces   1029
reflectance anisotropy spectroscopy
   (RAS)   1007
refractive index   478, 829, 946–972
 – elements   48
 – glasses   539, 543
 – polymers   478
 – Sellmeier dispersion formula   547
 – temperature dependence   548
refractive index n 619
 – boron compounds   610
 – cadmium compounds   684
 – gallium compounds   635
 – group IV semiconductors and
   IV–IV compounds   601–603
 – indium compounds   650
 – mercury compounds   691
 – zinc compounds   672
refractories

 – boride-based   452
 – carbide-based   458
 – nitride-based   468
 – oxide-based   438
 – silicide-based   472
refractory ceramics   437
refractory metals   303
 – alloys   303
      compositions   305
      dispersion-strengthened   304
 – annealing   311
 – chemical properties   308
 – crack growth behavior   325
 – creep elongation   316
 – creep properties   327
 – dynamic properties   318
 – evaporation rate   307
 – fatigue data   321
 – flow stress   316
 – fracture mechanics   322
 – grain boundaries   314
 – high-cycle fatigue properties   319
 – linear thermal expansion   306
 – low-cycle fatigue properties   320
 – mechanical properties   314
 – metal loss   308
 – microplasticity   318
 – oxidation behavior   308
 – physical properties   306
 – production routes   304
 – recrystallization   311
 – resistance against gaseous media
   309
 – resistance against metal melts
   309
 – specific electrical resistivity   307
 – specific heat   307
 – static mechanical properties   315
 – stress–strain curves   320
 – thermal conductivity   306
 – thermomechanical treatment   314
 – vapor pressure   307
 – Young's modulus   307
refractory metals alloys
 – application   306
 – products   306
refractory production
 – raw materials   444
relative permittivity   478
 – polymers   478
relaxation of semiconductors   991
relaxation of surface   986
 – metals   987
relaxor   906, 909, 918
remanent magnetization   922
remanent polarization   918

residual resistance ratio (RRR)   397
residual resistivity ratio (RRR)   338
resistivity
 – gallium compounds   630
response of material   46
Rf rutherfordium   94
Rh rhodium
 – elements   135
RHEED (reflection high-energy
   electron diffraction)   990
rhenium Re
 – elements   124
rhodium
 – alloys   386
 – applications   386
 – chemical properties   392
 – electrical properties   390
 – magnetic properties   391
 – mechanical properties   387
 – optical properties   392
 – phase diagrams   386
 – production   386
 – thermal properties   392
 – thermoelectrical properties   391
rhodium Rh
 – elements   135
ribbon silicates   433
RIE (reactive ion etching)   1061
Rn radon   128
Rochelle salt   904
Rochelle salt family   932
rod-like molecule   942
RT (room temperature)   49
RTP (room temperaure and standard
   pressure)   49
Ru ruthenium   131
rubidium Rb
 – elements   59
ruthenium Ru
 – alloys   399
 – applications   399
 – chemical properties   402
 – electrical properties   401
 – elements   131
 – lattice parameter   400
 – magnetic properties   401
 – mechanical properties   400
 – optical properties   402
 – phase diagrams   399
 – production   399
 – thermal properties   402
 – thermoelectric properties   401
rutherfordium Rf
 – elements   94
RW (weighted sound reduction)
   409

S

S sulfur 108
SAE (Society of Automotive Engineers) 221
SAM (self-assembled monolayer) 944
samarium Sm
– elements 142
SAW (surface acoustic wave) 912
Sb antimony 98
SbSI family 922
Sc scandium 84
$SC(NH_2)_2$ family 930
scandium Sc
– elements 84
scattering
– nanoscale objects 1048
scattering losses of a waveguide 1045
Schoenflies symbol 30
– elements 47
Schott AG 523
Schott code
– glasses 544
Schott filter glasses
– glasses 569
Schott glasses 8nnn 540, 541
SDR 997
Se selenium 108
seaborgium Sg
– elements 114
sealing glasses 527
– glasses 559
second
– SI base unit 14
secondary hardening 263
second-harmonic generation (SHG) 825, 906
second-order elastic constants see elastic constant 580
second-order phase transition 907
selenium Se
– elements 108
Sellmeier dispersion formula
– glasses 547
Sellmeier equations 826
SEM (scanning electron microscopy) 39
semiconductor 1003
– covalent 992
– field-effect mobility 1024
– III–V compounds 1004
– intrinsic Debye length 1021
– nanocrystal 1040
– polar 992
– quantum confinement 1036
– reconstruction 1004
– reconstruction model 991
– surface 990
– surface core level shift 1003
– surface Debye temperature 1017
– surface phonon 1017
– surface shift 1004
semiconductor band bending 1020
semiconductor nanostructures 1035
semiconductor surface
– Fermi level pinning 1025
– ionization energy 1003
semiconductors
– aluminium compounds 610
– boron compounds 604
– cadmium compounds 676
– chemical doping 576
– gallium compounds 621
– group IV semiconductors and IV–IV compounds 578–603
– III–V compounds 576, 604
– II–VI compounds 576, 652
– indium compounds 638
– introduction 575
– IV–IV compounds 576
– magnesium compounds 655
– mercury compounds 686
– oxides of Ca, Sr, and Ba 660
– physical properties 577
– table of contents 575
– zinc compounds 665
semi-solid metal processing (SSMP) 170
sensor
– chemiresistor-type 1043
SF1
– optical glasses 551
SF11
– optical glasses 551
SF2
– optical glasses 551
SF6
– glasses 537, 548
– optical glasses 551
SF66
– optical glasses 551
SFG (sum frequency generation) 825
SFM (superfluorinated material) 975
Sg seaborgium 114
shape memory 298
– nickel 279
shape-memory alloys
– TiNi 216
shear modulus 478
– elements 47
– polymers 478
shear rate 478
– polymers 478
SHG (second-harmonic generation) 825, 906
Shore hardness 478
– polymers 478
short pass filters
– glasses 566
short-range order 39
– glasses 524
Shubnikov groups
– crystallography 33
SI (Système International d'Unités) 3, 11
SI (the International System of Units) 12
SI base unit 13
SI definitions of magnetic susceptibility 48
SI derived units 16, 17
– with special names 17, 18
SI prefixes 19
Si silicon 88
SI units
– base quantities 13
– base units 13
Si_3N_4 ceramics 451
Si_3N_4 powders 472
SiC ceramics 451
side group 943
sievert
– SI unit of dose equivalent 19
silica
– glasses 524
silicate 433
silicate based glasses 526
silicide 473
– physical properties 472
silicon
– electromagnetic and optical properties 601
– electronic properties 589–594
– transport properties 598
silicon carbide
– band structure 590
– crystal structure, mechanical and thermal properties 578–588
– electromagnetic and optical properties 601

– electronic properties 589–594
– transport properties 595
silicon nitride 467
silicon Si
– crystal structure, mechanical and thermal properties 578–588
– elements 88
silicon steels
– grain-oriented 765
– non-oriented 763
silicon technology 1036
silicon-based lasers 1036
silicon–germanium alloys
– band structure 590
– transport properties 601
silicon-germanium alloys
– crystal structure, mechanical and thermal properties 578–588
– electromagnetic and optical properties 601
silicon–silicon oxide interface 1025
silver 330
– alloys 330
– application 330
– chemical properties 344
– crystal structures 333
– diffusion 342
– electrical properties 338
– intermetallic phases 333
– magnetic properties 339
– mechanical properties 335
– optical properties 341
– phase diagrams 331
– production 330
– ternary alloys 345
– thermal properties 340
– thermodynamic data 331
– thermoelectric properties 339
silver Ag
– elements 65
simple perovskite-type oxide 909
single hysteresis loop 904
SiO_2
– glasses 524
SK51
– optical glasses 551
Sm samarium
– elements 142
smectic C* phase 934
smectic phase 942
Sn tin 88
SNR (signal-to-noise ratio) 1060
soda lime glasses 528, 529, 534
sodium Na

– elements 59
soft annealing 224
soft magnetic alloys 758
– nanocrystalline 776
soft magnetic materials
– composite 759
– sintered 759
soft-mode spectroscopy 906
solar cell 1043
solder alloy 345
solder glasses 562
sol–gel synthesis
– nanostructured materials 1065
solid material
– structure 27
solid material, structure 27
solid surface energy 944
solid-state polymorphism 943
sorosilicate 433
sound velocity 478, 947–949, 956–962, 968–971
– elements 47
– polymers 478
source 1024
sp^3-bonded crystal 991
space charge function
– solid surfaces 1022
space charge layer
– semiconductor surface 1020
– solid surfaces 1020
space groups
– crystallography 31
SPARPES 1011
speed of light
– fundamental constant 13
spheroidal (nodular) graphite (SG) 268
spheroidite 223
spin accumulation 1049
spin diffusion length 1050
spin electronics
– applications 1057
– nanostructured materials 1031, 1049
spin polarization 1055
– nanostructured materials 1049
spin valve multilayers 1052
spin valve read head
– schematic diagram 1058
spin valve sensor 1053
spin-asymmetric material 1049
spinel structure
– crystallography 33
spin-electronic switch 1055
spin–orbit splitting energy Δ_{so}
– aluminium compounds 617

– cadmium compounds 681
– gallium compounds 628
– group IV semiconductors and IV–IV compounds 593
– indium compounds 644
– mercury compounds 689
– zinc compounds 670
spintronics
– nanostructured materials 1031, 1049
SPLEED 1011
spontaneous electric polarization 903
spontaneous polarization 917, 945
Sr strontium 68
$Sr_2Nb_2O_7$ family 920
SRI (sound reduction index) 409
$SrNb_2O_7$ family 909
$SrTeO_3$ family 919
$SrTiO_3$ 913
stabilized zirconia (PSZ) 448
stacking faults
– crystallography 41
stain resistance
– optical glasses 550
stainless steels 240
– austenitic 252
– duplex 257
– ferritic 246
– martensitic 250
– martensitic-ferritic 250
standard electrode potential
– elements 46
standard entropy
– elements 47
standard temperature and pressure (STP) 46
Stark effect
– nanostructured materials 1040
static dielectric constant 826
– elements 48
static dielectric constant 828–889
STC (sound transmission classification) 409
steam permeation 478
– polymers 478
steel
– austenitic 259
– carbon 227
– ferritic 258
– ferritic austenitic 259
– hardening 237
– heat-resistant 258, 261
– high-strength low-alloy (HSLA) 240

- low-alloy carbon steel 227
- mechanical properties 237
- stainless 240
- tool 262
stibiotantalite family 909
stiffness constant *see* elastic constant 580
STM (scanning tunneling microscopy) 988, 992, 997
STM spectroscopy 1005
STN (supertwisted nematic) effect 944
STO (SrTiO$_3$) 1056
storage capacity of hard disks 1048
storage density evolution of hard disk drives 1048
storage media 1060
- arrays of nanometer-scale dots 1060
- limits 1060
- technology 1060
storing information on the sidewalls of the dots 1062
strain
- polymers 478
strength
- glasses 534
stress 478
- polymers 478
stress at 50% strain (elongation) 478
- polymers 478
stress at fracture 478
stress at yield 478
- polymers 478
stress birefringence
- glasses 539, 549
stress intensity factor
- glasses 534
strong-confinement regime
- nanostructured materials 1037, 1038
strontium oxide
- crystal structure, mechanical and thermal properties 660
- electromagnetic and optical properties 664
- electronic properties 661
- transport properties 663
strontium Sr
- elements 68
strontium titanate 913
structural parameters 990
structural phase transitions 906
structure

- diamond-like 979
structure type
- crystallography 33
Strukturbericht type 47
sublattice 903
sublattice polarization 904
submicrometer magnetic dots 1060
substituted mesogens (liquid crystals)
- physical properties 968
sulfur S
- elements 108
sum frequency generation (SFG) 825
superalloys
- Ni-based cast 288
- nickel 294
superconducting high-T_c
- crystal structure 712
superconducting oxides
- high-T_c chemical composition 712
superconductivity
- elements 48
superconductor 695
- borides 745
- borocarbides 746, 747
- carbides 745
- commercial Nb$_3$Sn 709
- critical temperature 699
- crystal structure 712
- Debye temperature 696
- device applications 719
- high-T_c cuprates 712, 713, 720
- industrial wire performance 719
- metallic 696
- Nb alloys 702
- non-metallic 712
- Pb alloys 696
- pinning 717
- practical metallic 704
- production Nb$_3$Sn 708
- Sommerfeld constant 696
- SQUIDs 720
- structural data 720
- thermodynamic properties 696
- Type I 695
- Type II 695
- V alloys 700
- vortex lines 717
- Y–Ba–Cu–O 723
supercooled liquid
- glasses 524
supercooled mesophase 945
superstructures

- crystallography 41
supertwisted nematic (STN) effect 944
Supremax
- glasses 527
surface
- Curie temperature 1009
- diagram 979
- ionization energy 1003
- magnetic 1008
- semiconductor 990
- structure of an ideal 979
surface band structure 996
surface Brillouin zone (SBZ) 996
surface conductivity
- solid surfaces 1024
surface core level shifts (SCLS) 998
- solid surfaces 1003, 1004
surface differential reflectivity (SDR) 1008
surface excess conductivity 1024
surface magnetization 1011
surface mobility 1024
surface of diamond 1004
surface phonon 1012
- dispersion 1019
- metal 1013
- mode 1012
surface plasmon
- absorption of nanoparticles 1046
- dispersion curve 1001
surface resistivity 478
- polymers 478
surface resonance 996
- phonons 1012
surface response
- dielectric theory 1007
surface state
- acceptor 1022
- band 1005
- donor 1022
- transitions 1005
surface state bands
- solid surfaces 1004
surface states 996
surface tension 948–950, 955–962, 968–971, 975
- elements 47
surface tension (γ_{LV}) 943
surfaces 979
surgical implant alloys
- cobalt-based 277
susceptibility

– magnetic   48
– mass   48
– molar   48
– nonlinear dielectric   825, 829
– second-order nonlinear dielectric   826
– third-order nonlinear dielectric   826
SV (spin valve)   1052
symmetry elements of point groups   30
synthesis of clusters
– gas-phase production   1066
– nanostructured materials   1065
synthetic silica
– glasses   557

T

T tritium   54
TA
– phonon spectra   915
Ta tantalum   105
Ta-based alloys   318
tailoring of the electronic wave function   1032
tantalum Ta
– elements   105
Tb terbium   142
TC 12 (Technical Committee 12 of ISO)   12
Tc technetium   124
Te tellurium   108
technetium Tc
– elements   124
technical ceramics   437
technical coppers   297
technical glasses   527, 530
technical specialty glasses   526
tellurium Te
– elements   108
TEM (transmission electron microscopy)   39
TEM image of a superlattice of Au clusters   1066
TEM views of single dots   1062
temper graphite (TG)   268
temperature dependence of carrier concentration
– group IV semiconductors and IV–IV compounds   596
temperature dependence of electrical conductivity
– indium compounds   647
temperature dependence of electronic mobilities

– indium compounds   649
temperature dependence of energy gap
– indium compounds   645
temperature dependence of linear thermal expansion coefficient
– cadmium compounds   677
– magnesium compounds   656
temperature dependence of the lattice parameters
– group IV semiconductors and IV–IV compounds   580–582
temperature dependence of thermal conductivity
– group IV semiconductors and IV–IV compounds   599, 600
– indium compounds   650
temperatures of phase transitions   946–976
tempering of steel   223
template synthesis   944
tensile strength
– elements   47
tensor
– elastooptic   826
– piezoelectric strain   826
terbium Tb
– elements   142
terminal group   943
ternary alloys   298
terne steel coatings   415
TFT (thin-film transistor)   944
TGS family   932
Th thorium   151
thallium Tl
– elements   78
thermal and thermodynamic properties
– elements   46
thermal conductivity κ 478, 945, 947–949, 956, 958
– aluminium compounds   618
– cadmium compounds   682
– elements   46, 47
– gallium compounds   634
– group IV semiconductors and IV–IV compounds   599
– indium compounds   650
– magnesium compounds   659
– mercury compounds   690
– oxides of Ca, Sr, and Ba   663
– polymers   478
– zinc compounds   670
thermal expansion
– glasses   526
thermal expansion coefficient, linear

– elements   47
thermal gap
– indium compounds   645
thermal properties
– group IV semiconductors   578–588
– III–V compounds   610, 621, 638
– III–V semiconductors   604
– II–VI compounds   652, 655, 660, 665, 676, 686
– IV–IV compound semiconductors   578–588
– technical glasses   533, 536
thermal vibrations
– surface phonons   1012
thermal work function
– elements   48
thermally activated flux flow (TAFF)   718
thermochromic material   944
thermodynamic properties
– elements   47
thermoelectric coefficient
– elements   48
thermoelectric power
– oxides of Ca, Sr, and Ba   664
thermography   941, 944
thermomechanical treatment (TMT)   314
thermosets   481, 512
– diallyl phthalate (DAP)   512, 514
– epoxy resin (EP)   514, 515
– melamine formaldehyde (MF)   512, 513
– phenol formaldehyde (PF)   512, 513
– polymers   512
– silicone resin (SI)   514, 515
– unsaturated polyester (UP)   512–514
– urea formaldehyde (UF)   512, 513
thermotropic liquid crystal   942
thin film   906
thin-film transistor (TFT)   944
thixomolding   170
thorium Th
– elements   151
three and four-ring systems
– liquid crystals   964
three-dimensional long-range order   941
three-ring system   964
three-wave interactions
– in crystals   825
thulium Tm
– elements   142

Ti titanium 94
time-temperature-transformation (TTT) diagram 238
tin Sn
– elements 88
titanates 450
titanium 206
– commercially pure grades 207
– creep behavior 208
– creep strength 210
– hardness 207
– high-temperature phase 206
– intermetallic materials 210
– phase transformation 206
– sponge 207
– superalloys 210
– titanium alloys 206
titanium alloys 209
– applications 209
– chemical composition 209
– chemical properties 213
– mechancal properties 213
– mechanical properties 209
– physical properties 213
– polycrystalline 213
– single crystalline 213
– thermal expansion coefficient 214
titanium dioxide
– mechanical properties 450
– thermal properties 450
titanium oxide
– phase diagram 206
titanium Ti
– elements 94
Tl thallium 78
Tm thulium 142
TMR (tunnel magnetoresistance) 1054, 1055
TN (twisted nematic)
– liquid crystals 944
TO 915
tool steels 262
torsional modulus
– optical glasses 550
total losses 763
transformation temperature
– glasses 524
transition range
– glasses 524
transition temperature
– glasses 525
transitions
– surface states 1005
transmission spectra

– colored glasses 566, 567
transmission window
– glasses 524
transmittance
– glasses 548
transmittance of glasses
– color code 549
transport properties
– group IV semiconductors and IV–IV compounds 595–601
– III–V compounds 608, 617, 629, 647
– II–VI compounds 655, 659, 663, 670, 682, 689
transverse acoustic branch 915
transverse optical branch 915
transverse optical mode 906
triple point of water 48
tritium T
– elements 54
truncated crystal 986
tungsten bronze-type family 909, 920
tungsten W
– elements 114
tunnel junction
– magnetic 1053
tunnel magnetoresistance
– function of field and temperature 1057
tunnel magnetoresistance as a function of magnetic field 1055
tunneling
– nanostructured materials 1053
tunneling mechanism
– nanostructured materials 1043
twisted nematic (TN) effect 944
two-dimensional liquid 942
two-photon absorption coefficient 829
two-ring systems with bridges
– liquid crystals 955
two-ring systems without bridges
– liquid crystals 947
Type II superconductors
– anisotropy coefficients 716
– coherence lengths 716
– high-T_c cuprate compounds 716
type metals 414

U

U uranium 151
ultrahigh density storage media 1049, 1060

unalloyed coppers 296
uniaxial crystals 826
Unified Numbering System for Metals and Alloys (UNS) 296
unit cell of Si(111) 7×7 995
units
– amount of substance 14
– atomic 21
– atomic units (a.u.) 22
– candela 15
– CGS units 21
– coherent set of 20
– crystallography 21
– electric current 14
– general tables 4
– length 13
– luminous intensity 15
– mass 14
– natural 21
– natural units (n.u.) 21
– non-SI 22
– non-SI units 20, 21
– other non-SI units 23
– temperature 14
– the international system of 11
– time 14
– used with the SI 20
– X-ray-related units 22
units of physical quantities
– fundamental constants 3
units outside the SI 20
UNS (Unified Numbering System) 221
UPS 998
uranium U
– elements 151
UTS – ultimate tensile strength 219

V

V vanadium 105
van der Waals attraction 1019
vanadium V
– elements 105
vertical nanomagnets 1062
vertical relaxation of metals 989
VFT (Vogel, Fulcher, Tammann) equation
– glasses 533
Vicat softening temperature 477
– polymers 477
Vickers hardness
– elements 47

vinylpolymers 480, 489–492
– poly(acrylonitrile-co-butadiene-co-styrene) (ABS) 492–494
– poly(acrylonitrile-co-styrene-co-acrylester) (ASA) 492–494
– poly(styrene-co-acrylnitrile) (SAN) 489, 490, 492
– poly(styrene-co-butadiene) (SB) 489–491
– poly(vinyl carbazole) (PVK) 492, 493
– polystyrene (PS) 489–491
VIP (viewing-independent panel) 975
viscosity 478, 948, 950–954, 963–965, 967, 968, 975
– dynamic 943
– elements 47
– glasses 524
– kinematic 943
– optical glasses 556
– polymers 478
– technical glasses 533
– temperature dependence 525
viscosity of glasses
– temperature dependence 534
vitreous silica
– electrical properties 557
– gas solubility 557
– glasses 526, 556
– molecular diffusion 557
– optical constants 557
vitreous solder glasses 562
Vitronit
– glasses 559
volume compressibility
– elements 47
volume magnetization
– elements 48
volume of primitive cell
– crystallographic formulas 986
volume resistivity 478
– polymers 478
volume–temperature dependence
– glasses 524
Vycor^{TM}
– glasses 527

W

W tungsten 114
wavelength dependence of refractive index n
– indium compounds 651

WDX (wavelength-dispersive analysis of X-rays) 39
weak-confinement regime
– nanostructured materials 1037, 1038
wear-induced surface defects
– glasses 535
Weibull distribution
– glasses 535
weight fraction
– glasses 527
Wood's metal 420
work function Φ
– metal 997
– solid surfaces 997
work hardening wrought copper alloys 300
wrought alloys 298
wrought magnesium alloys 164
wrought superalloys 284
wtppm (weight part per million) 407
Wyckoff position
– crystallography 32

X

Xe xenon 128
xenon Xe
– elements 128
X-ray diffraction 39
X-ray interferences
– crystallography 27

Y

Y yttrium 84
Yb ytterbium 142
Young's modulus 823
– elements 47
– optical glasses 550
YS – yield stress 219
ytterbium Yb
– elements 142
yttrium Y
– elements 84
Y–Ba–Cu–O
– critical current density 733
– crystal defects 728
– crystal structure 723
– electric resistivity 730
– grain boundaries 730
– hole concentration 733
– lattice parameters 726
– lower critical field 734
– oxygen content 724

– pinning 735
– substitutions 725
– superconducting properties 731
– thermal conductivity 730
– transition temperature 733
– upper critical field 734

Z

zeolites
– nanostructured materials 1031, 1065
Zerodur$^®$
– glasses 558
– linear thermal expansion 558
zinc compounds
– crystal structure, mechanical and thermal properties 665
– effective hole mass 670
– electromagnetic and optical properties 672
– electronic properties 668
– mechanical and thermal properties 665
– optical properties 672
– thermal properties 665
– transport properties 670
zinc oxide
– crystal structure, mechanical and thermal properties 665
– electromagnetic and optical properties 672
– electronic properties 667
– transport properties 670
zinc selenide
– crystal structure, mechanical and thermal properties 665
– electromagnetic and optical properties 672
– electronic properties 667
– transport properties 670
zinc sulfide
– crystal structure, mechanical and thermal properties 665
– electromagnetic and optical properties 672
– electronic properties 667
– transport properties 670
zinc telluride
– crystal structure, mechanical and thermal properties 665
– electromagnetic and optical properties 672
– electronic properties 667
– transport properties 670

zinc Zn
– elements 73
zircaloy 219
– irradiation effect 219
zirconium
– alloys 217
– bulk glassy alloys 218
– bulk glassy behavior 220
– low alloy materials 217
– nuclear applications 218
– technically-pure materials 217
zirconium dioxide 448

zirconium Zr
– elements 94
ZLI-1132
– liquid crystals 975
Zn zinc 73
Zr zirconium 94

Periodic Table of the Elements

Legend:
- 18 / VIIIA — IUPAC Notation / CAS Notation
- 2 / He — Atomic Number / Element Symbol
- Shaded: Unstable Nuclei

Main Groups

1 IA	2 IIA	13 IIIA	14 IVA	15 VA	16 VIA	17 VIIA	18 VIIIA	Shells
1 H							2 He	K
3 Li	4 Be	5 B	6 C	7 N	8 O	9 F	10 Ne	K–L
11 Na	12 Mg	13 Al	14 Si	15 P	16 S	17 Cl	18 Ar	K–L–M
19 K	20 Ca	31 Ga	32 Ge	33 As	34 Se	35 Br	36 Kr	–L–M–N
37 Rb	38 Sr	49 In	50 Sn	51 Sb	52 Te	53 I	54 Xe	–M–N–O
55 Cs	56 Ba	81 Tl	82 Pb	83 Bi	84 Po	85 At	86 Rn	–N–O–P
87 Fr	88 Ra							–O–P–Q

Subgroups

3 IIIB	4 IVB	5 VB	6 VIB	7 VIIB	8 VIII(1)	9 VIII(2)	10 VIII(3)	11 IB	12 IIB	Shells
21 Sc	22 Ti	23 V	24 Cr	25 Mn	26 Fe	27 Co	28 Ni	29 Cu	30 Zn	–L–M–N
39 Y	40 Zr	41 Nb	42 Mo	43 Tc	44 Ru	45 Rh	46 Pd	47 Ag	48 Cd	–M–N–O
57 La	72 Hf	73 Ta	74 W	75 Re	76 Os	77 Ir	78 Pt	79 Au	80 Hg	–N–O–P
89 Ac	104 Rf	105 Db	106 Sg	107 Bh	108 Hs	109 Mt	110 Ds	111 Rg	112	–O–P–Q

Lanthanides (Shells –N–O–P)

58 Ce	59 Pr	60 Nd	61 Pm	62 Sm	63 Eu	64 Gd	65 Tb	66 Dy	67 Ho	68 Er	69 Tm	70 Yb	71 Lu

Actinides (Shells –O–P–Q)

90 Th	91 Pa	92 U	93 Np	94 Pu	95 Am	96 Cm	97 Bk	98 Cf	99 Es	100 Fm	101 Md	102 No	103 Lr

Most Frequently Used Fundamental Constants

CODATA Recommended Values of Fundamental Constants

Quantity	Symbol and relation	Numerical value	Unit	Relative standard uncertainty
Speed of light in vacuum	c	299 792 458	m/s	Fixed by definition
Magnetic constant	$\mu_0 = 4\pi \times 10^{-7}$	$12.566370614\ldots \times 10^{-7}$	N/A^2	Fixed by definition
Electric constant	$\varepsilon_0 = 1/(\mu_0 c^2)$	$8.854187817\ldots \times 10^{-12}$	F/m	Fixed by definition
Newtonian constant of gravitation	G	$6.6742(10) \times 10^{-11}$	m^3/(kg s^2)	1.5×10^{-4}
Planck constant	h	$4.13566743(35) \times 10^{-15}$	eV s	8.5×10^{-8}
Reduced Planck constant	$\hbar = h/2\pi$	$6.58211915(56) \times 10^{-16}$	eV s	8.5×10^{-8}
Elementary charge	e	$1.60217653(14) \times 10^{-19}$	C	8.5×10^{-8}
Fine-structure constant	$\alpha = (1/4\pi\varepsilon_0)(e^2/\hbar c)$	$7.297352568(24) \times 10^{-3}$		3.3×10^{-9}
Magnetic flux quantum	$\Phi_0 = h/2e$	$2.06783372(18) \times 10^{-15}$	Wb	8.5×10^{-8}
Conductance quantum	$G_0 = 2e^2/h$	$7.748091733(26) \times 10^{-5}$	S	3.3×10^{-9}
Rydberg constant	$R_\infty = \alpha^2 m_e c/2h$	10 973 731.568525(73)	1/m	6.6×10^{-12}
Electron mass	m_e	$9.1093826(16) \times 10^{-31}$	kg	1.7×10^{-7}
Proton mass	m_p	$1.67262171(29) \times 10^{-27}$	kg	1.7×10^{-7}
Proton–electron mass ratio	m_p/m_e	1836.15267261(85)		4.6×10^{-10}
Avogadro number	N_A, L	$6.0221415(10) \times 10^{23}$		1.7×10^{-7}
Faraday constant	$F = N_A e$	96 485.3383(83)	C	8.6×10^{-8}
Molar gas constant	R	8.314472(15)	J/K	1.7×10^{-6}
Boltzmann constant	$k = R/N_A$	$1.3806505(24) \times 10^{-23}$ $8.617343(15) \times 10^{-5}$	J/K eV/K	1.8×10^{-6} 1.8×10^{-6}
Josephson constant	$K_J = 2e/h$	$483 597.879(41) \times 10^9$	Hz/V	8.5×10^{-8}
von Klitzing constant	$R_K = h/e^2 = \mu_0 c/2\alpha$	25 812.807449(86)	Ω	3.3×10^{-9}
Bohr magneton	$\mu_B = e\hbar/2m_e$	$927.400949(80) \times 10^{-26}$ $5.788381804(39) \times 10^{-5}$	J/T eV/T	8.6×10^{-8} 6.7×10^{-9}
Atomic mass constant	$u = (1/12)m(^{12}C)$ $= (1/N_A) \times 10^{-3}$ kg	$1.66053886(28) \times 10^{-27}$	kg	1.7×10^{-7}
Bohr radius	$a_0 = \alpha/4\pi R_\infty$ $= 4\pi\varepsilon_0 \hbar^2/m_e e^2$	$0.5291772108(18) \times 10^{-10}$	m	3.3×10^{-9}
Quantum of circulation	$h/2m_e$	$3.636947550(24) \times 10^{-4}$	m^2/s	6.7×10^{-9}